I0058425

# Methodologies for Soil and Sediment Fractionation Studies

## Single and Sequential Extraction Procedures

*This book is dedicated to Ben Griepink who initiated the EC pro-gramme on extractable trace elements in soils and sediments, opening a way to ensure the comparability of data for operationally-defined measurements.*

# Methodologies in Soil and Sediment Fractionation Studies
## Single and Sequential Extraction Procedures

Edited by

**Ph. Quevauviller**
*European Commission, DG Research, Brussels, Belgium*

RS•C

**ROYAL SOCIETY OF CHEMISTRY**

ISBN 0-85404-453-1

A catalogue record for this book is available from the British Library

© The Royal Society of Chemistry 2002

*All Rights Reserved*

*Apart from any fair dealing for the purposes of research or private study, or criticism or review as permitted under the terms of the UK Copyright, Designs and Patents Act, 1988, this publication may not be reproduced, stored or transmitted, in any form or by any means, without the prior permission in writing of The Royal Society of Chemistry, or in the case of reprographic reproduction only in accordance with the terms of the licences issued by the Copyright Licensing Agency in the UK, or in accordance with the terms of the licences issued by the appropriate Reproduction Rights Organization outside the UK. Enquiries concerning reproduction outside the terms stated here should be sent to The Royal Society of Chemistry at the address printed on this page.*

Published by The Royal Society of Chemistry
Thomas Graham House, Science Park, Milton Road, Cambridge CB4 0WF, UK

Registered Charity Number 207890

For further information see our web site at www.rsc.org

Typeset by Vision Typesetting, Manchester, UK
Printed and bound by MPG Books Ltd, Bodmin, Cornwall, UK

# *Foreword*

During the last thirty years, the emphasis of analysis techniques for the evaluation of environmental impact of polluted solid materials has changed from a simple determination of total concentrations towards a more sophisticated fractionation of the various carrier matrices with the bulk soil, sediment or solid waste material sample. This change in focus results from a recognition that biogeochemical and especially the ecotoxicological significance of a given pollutant input is determined by its specific binding form and coupled reactivity rather than by its accumulation rate in the solid material. In a broad sense, *speciation* has been defined as either the *process of identifying and quantifying* the different, defined species, forms or phases present in a material, or the *description of the amounts and kinds* of these species, forms or phases present.

Such information is – as outlined by Allan Ure and Christine Davidson in their *Chemical Speciation in the Environment* – of relevance to scientists with many different backgrounds: chemists, biologists, soil and sediment scientists, physicists, and specialists in various aspects of nutrition and medicine. In some instances, however, the quality of scientific accomplishment could not keep up with the attraction of the concept. Regarding sequential extraction, some reviewers came to conclusions as pessimistic as 'these techniques represent nothing but an operational tool and complementary approach until physical techniques be available with the new generation of microprobes and other sophisticated instruments'. Following a subtle discussion, the International Union of Pure and Applied Chemistry (IUPAC) insisted that the term 'speciation' should not be applied to operationally-defined procedures such as single and sequential extraction methods.

On the other hand, this was and still is exactly the type of information required by the practitioners. What they need is a cost-effective and reliable characterisation of at least some of the most important forms of an element or trace organic compound in order to understand the transformations between forms which can occur and to infer from such information the likely environmental consequences. It is clear that a significant economic impact will arise from this approach. The differentiating view may enable more comprehensive interpretation regarding the potential long-term risks and will result in more adequate administrative priorities, political decisions, management schemes and technical performance.

This refers to the traditional applications such as dredging operations, treatment of solid waste materials, installation of barriers, *etc.*, but also to new areas, such as utilisation of nature-near remediation processes for certain types of contaminated sites and, even more challenging, future efforts for evaluating the so-called 'brown fields', *i.e.* recycling of abandoned land. In both latter aspects a science-based approach, involving differentiation of hazards from typical pollutant species and respecting processes of natural attenuation, could save expenditures well in the range of a billion euros or dollars.

Extraction procedures, either single or sequential, may be considered as a typical child of pragmatic American soil and sediment researchers. It is now growing particularly well under the climate of the emerging European Union, with its inherent enthusiasm for scientific collaboration and the external need to channelise their results. During the last twenty years, we have experienced not only an increase in scientific quality by the coordinated research projects but also an even more remarkable progress in the administration of technology development and standardisation. An example of a successful outcome of such research activities is the development of the pH dependence test as a tool for environmental risk assessment studies: this test is now standardised at the CEN level (European Committee for Standardisation) and encompasses both single and sequential extraction methods into a scheme that allows more direct relationships to be made with true speciation as derived from geochemical speciation models.

Characteristic of the latter are the efforts of the Measurement and Testing programme (formerly BCR) of the European Commission to improve the quality of analysis in speciation studies. These have taken the form of interlaboratory studies in several fields of speciation by chemical analysis. These interlaboratory studies aimed to (i) minimise errors in sampling, sample treatment and analysis, (ii) identify the most appropriate analytical procedures, and (iii) harmonise the total analytical procedure into an agreed analytical procedural protocol that could be used for the preparation of reference materials certified for species contents. For example, in the process of evolving such reference materials for the validation of the speciation of heavy metals in soils by selective extraction and in sediments by sequential extraction, under the auspices of BCR, some 30 expert European laboratories took part and developed the selective extraction protocols and the simplified three-stage selective procedure used. Meanwhile, the certified soil and sediment materials are available commercially from BCR. The need for reference materials certified for species contents is now well recognised and a number are in preparation by a number of different bodies.

The present book, edited by the project manager for BCR, includes both theoretically and practically important background information to this ambitious programme of the Commission of the European Community. The authors of the individual contributions are not only leading experts in their field, but have also accompanied the preparation of certified reference materials (CRMs) as quality control tools during the last twenty years, involving intensive contacts with both the world-wide scientific community and with practioners in quite different fields of application. In this way, the present compilation with its 'innovative' features related to harmonisation, standardisation and production of CRMs, will become an invaluable source of reference for the still growing field of fractionation in soils, sediments and other environmentally-relevant solid materials.

Ulrich Förstner
Hamburg, November 2001

# Preface

Environmental studies on soil and sediment analysis are often based on the use of leaching or extraction procedures (*e.g.* single or sequential extraction procedures), enabling broader forms or phases to be measured (*e.g.* 'bioavailable' forms of elements), which are in most instances sufficient for the purpose of environmental policy. The development and application of extraction schemes started at the end of the 1970s for the purpose of studying the availability of heavy metals (and thus estimating the related phytotoxic effects) and their mobility (*e.g.* from a soil and potential groundwater contamination). This approach has long been referred to as 'speciation' but, according to the recent IUPAC definition, the term 'fractionation' is now preferred. One would have thought that the lack of uniformity of these schemes would have stopped their use after some years. However, they have been adapted to various case studies and are still widely used for soil and sediment studies, as reflected by the number of scientific papers that are published on the subject. Similar approaches have been applied to the study of phosphorus release from lake sediments in relation to eutrophication studies. More recently, extraction and leaching procedures have been evaluated and applied to the determination of trace organic compounds in contaminated soils and wastes. The question of comparability of data obtained from various schemes and different laboratories has been raised at several occasions. The European Commission (through the BCR programme and its successors) has been the only organisation that tried to systematically evaluate this comparability (through the organisation of interlaboratory studies) and to propose a harmonisation of the most frequently used approach. This resulted in the production of Certified Reference Materials for extractable heavy metals and phosphorus forms in soils and sediments. Efforts were also made, through networking activities, to apply existing extraction/leaching schemes to studies of organic compounds in contaminated soils and wastes.

This book gives an account of the work carried out under the auspices of the European Commission and applications by research laboratories. The first chapter sets the scene by recalling the harmonisation needs for operationally-defined measurements. The second chapter describes in details the programme for adopting a common sequential extraction scheme for sediment analysis and certifying reference materials. A similar approach is described in the third chapter, dealing with the harmonisation of single extraction schemes for soil analysis along with the production of CRMs. The fourth chapter gives an overall picture of the application of extraction schemes to industrially-contaminated soils for heavy metal studies. In the fifth chapter, extraction methodologies are also thoroughly studied for harmonisation purposes, but the described programme focuses on phosphorus forms in lake sediment. The sixth chapter presents recent results on the application of extraction schemes to the determination of trace organic compounds in soil matrices. Finally, the seventh chapter describes discussions held within a thematic network aiming to harmonise leaching/extraction schemes in a multidisciplinary fashion (*i.e.* involving experts from various disciplines), with a focus on leaching methodologies applied to waste analyses.

This book has been written by experienced practitioners. It aims to serve as a practical reference for environmental chemists (and postgraduate students) who need in-depth information on the use of operationally-defined procedures for soil and sediment studies. The critical discussions of the methods makes it unique in this respect.

The editor gratefully acknowledges the authors for their time and motivation in preparing their contributions. A special thank you goes to all the laboratories which have participated in the various collaborative studies (listed at the end), without which this volume would not have been possible.

Philippe Quevauviller
Brussels, December 2001

# Contents

CHAPTER 1

# SM&T Activities in Support of Standardisation of Operationally-defined Extraction Procedures for Soil and Sediment Analysis

Ph. QUEVAUVILLER

European Commission, DG Research, Brussels, Belgium

## 1.1 Introduction

The environmental ecotoxicity and mobility of heavy metals is strongly dependent upon their specific chemical forms or way of binding. Consequently, their toxic effects and biogeochemical pathways can only be studied on the basis of the determination of these forms. The determination of chemical species (*e.g.* organometallic compounds) is often difficult in soil and sediment matrices and, to date, only very few compounds have been reported to be accurately determined in sediment (*e.g.* tributyltin, methylmercury),[1,2] using *e.g.* hyphenated techniques involving a succession of analytical steps (extraction, separation, detection). In practice, environmental studies on soil and sediment analysis are often based on the use of leaching or extraction procedures (*e.g.* single or sequential extraction procedures), enabling broader forms or phases to be measured (*e.g.* 'bioavailable' forms of elements), which are in most instances sufficient for the purpose of environmental policy.[3,4] This type of determination is often referred to as 'speciation' although, strictly speaking (see the recent IUPAC definition of 'speciation',[5]) this term should not be applied to operationally-defined procedures. Speciation would cover the determination of well-defined chemical species (*e.g.* organometallic compounds, metals with different oxidation states *etc.*), whereas the extracted 'forms' should be only related to the extractant used, *e.g.* EDTA-extractable element, and not as *e.g.* 'bioavailable', 'mobile' *etc.* forms, which are interpretations of data rather than results of actual measurements. This type of measurement is also referred to as 'fractionation'.

For heavy metals, the development and use of extraction schemes started at

1

the end of the 1970s and aimed to evaluate the metal fractions available to plants (and thus estimate the related phytotoxic effects) and the environmentally access-ible trace metals (*e.g.* mobility of metals from a soil and potential groundwater contamination);[6-8] these schemes have been adapted and are still widely used for soil and sediment studies as reflected by the number of recently published papers dealing with their applications to various environmental studies.[9-12] Similarly, extraction schemes were developed for studying the release of phosphorus from lake sediment in relation to eutrophication studies.[13] Finally, operationally-defined procedures are increasingly considered for environmental studies related to the mobility of trace organic compounds[14,15] which are strongly matrix-dependent and require complex procedures involving different analytical steps (extraction, clean-up, separation, detection).

Besides the usefulness of these schemes, however, it was recognised that the lack of uniformity in the procedures used did not allow the results to be compared world-wide nor the methods to be validated. Indeed, the results obtained are 'operationally-defined' which means that the 'forms' of pollutants are defined by the determination of extractable contents using a given procedure and that, therefore, the significance of the analytical results is highly dependent on the extraction procedures used.

Results are useful and usable only if they correspond to well-defined and accepted procedures. In other words, the only means for achieving sound inter-pretation and basis for decisions is to achieve comparability of results, which is closely linked to a consensus with respect to the used procedures, followed by their validation, and their possible implementation as a standard.

This introductory chapter describes an approach followed by the Standards, Measurements and Testing programme (formerly BCR) of the European Com-mission for harmonising single and sequential extraction procedures for soil and sediment analyses, with the aim to provide laboratories with reference schemes that could later become international standards. Details on the different selected and tested schemes for trace metals (EDTA, DTPA, acetic acid for soil analysis and three-step sequential extraction scheme for sediment analysis) and phos-phorus forms (sequential extraction scheme) are given in Chapters 2–4 of this book. As discussed later, the collaborative testing of these schemes, along with the preparation of related Certified Reference Materials (CRMs), has a clear effect on their world-wide use which is increasingly reflected by the litera-ture.[16-21] The schemes are not standardised *sensu stricto* (*i.e.* they were not adopted as official standards by an international standardisation organisation) but they fulfil the same role in enabling data comparability in this analytical field.

## 1.2   Standardisation

Many discussions have arisen on the risks that standardisation might 'fossilise' progress in analytical science in a wide range of cases. It is, however, generally accepted that the only way to achieve comparability when using operationally-defined procedures is to standardise them and apply them, following very strictly the written protocols. This does not mean that improvements should not be investigated to ensure progress in the use and result interpretation of these schemes; in this case standardisation offers scientists a possibility to speak the same language and decision-makers a way to identify better possible strategies for environmental risk assessment. Extraction tests are widely used for the assessment of the release of inorganic contaminants from soils, sludges and sediments. In many instances, these schemes are included in national (or sometimes regional) regulations. The International Standardization Organization (ISO) is coordinating working groups on soil quality (*e.g.* ISO TC/190) with the aim to identify a range of tests that would be acceptable for possible standardisation. Expert consultations and discussions are based on the selection of existing extraction schemes, *e.g.* EDTA, DTPA, calcium chloride *etc.* which have to be first accepted as candidate standard tests (on the basis of their scientific significance), then demonstrated to be applicable to various matrices (easiness of use, ruggedness) and possibly tested by expert organisations. This approach requires extensive consultations and possibly interlaboratory testing of the selected candidate standard procedures.

One of the principles of standardisation is to 'write what is done' and to 'do what is written'. A standard may be defined as a reference text, which has been elaborated by a recognised – national or international – organisation (*e.g.*

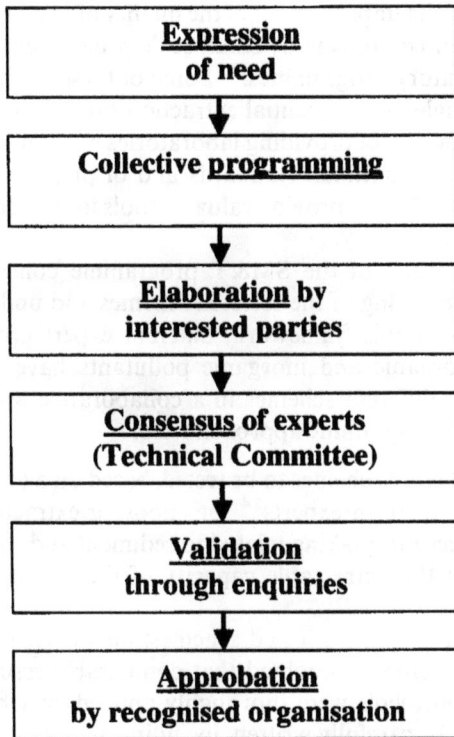

**Figure 1**  *Preparation process of a standard (adapted after ref. 22)*

AENOR in Spain, AFNOR in France, DIN in Germany, BS in the United Kingdom *etc.*, CEN for the European Union and ISO at the world scale) after agreement of all interested parties. Figure 1 recalls the process of preparation of a standard.[22]

Support to standardisation is achieved through the funding and organisation of interlaboratory studies, which aim to demonstrate the applicability of standards and to establish minimum technical requirements. Collaborative testing may indeed help standardisation organisations to take the decision to adopt or not a given procedure as an international standard. The approach followed by the SM&T programme and its successor (the European Commission) is a good example of such supporting activity. It should be stressed that the SM&T programme funded research and feasibility studies (including interlaboratory testing) but was not responsible for the adoption of standards nor for their implementation. This type of RTD programme should hence be seen as a tool for standardisation bodies but not as standardisation organisation itself.

## 1.3  SM&T Activities for the Harmonisation of Extraction Schemes

The SM&T programme (also referred to as BCR throughout the text) has often launched projects for the improvement of the quality of analytical measurements in a stepwise manner, *i.e.* starting by small-scale projects and developing them into wide interlaboratory programmes.[23] Some of these projects dealt with the harmonisation of single and sequential extraction procedures for soil and sediment analysis with the aim of providing laboratories with reference schemes that could later become international standards and of preparing related certified reference materials (CRMs) to provide valuable tools for validating methods and for quality control.

The stepwise approach of the SM&T programme consisted of feasibility studies, collaborative testing of the selected schemes and undertaking certification campaigns. Under this framework, different expert groups on soils and sediments in both organic and inorganic pollutants have been created and worked out together different schemes in a collaborative way. The steps considered in the SM&T programme approach were:

- Selection of the different schemes to be tested, based on a literature search and consultations of European experts[24] for choosing extractants which led to good results for decision-making related to sediment and soil analyses, taking into consideration the extractable capacity of the agent in relation to the studied matrices.
- Collaborative testing of the selected schemes: the analysis of the results obtained in these exercises highlighted that comparable results were only obtained when the protocols were thoroughly applied, which implied that the procedures had to be carefully written, including all operational details. Calibration errors were actually found to be the main sources of error in both single and sequential extraction schemes.[4,24,25]

- Appropriate selection of the test and candidate certified samples: in particular complying with the requirements of homogeneity and stability, representativeness of the studied matrices (calcareous, siliceous, organic, *etc.*) and with measurable analyte contents (*i.e.* not too close to the detection limits of the selected analytical techniques). For both single and sequential extraction schemes, interlaboratory studies were designed and conducted with soil and sediment reference materials originating from the Joint Research Centre of Ispra, Italy.[24]
- Validation of extraction methods: the concept of method validation in the case of operationally-defined procedures may be understood by the acceptance and testing of common schemes, possibly proposed as standardised methods to official standardisation organisations. Validation can be carried out as described above, *i.e.* through the organisation of interlaboratory studies in which a group of laboratories receive 'real case' samples to be analysed following a strict protocol. The experience has shown that, while the tests carried out in some schemes resulted in a satisfactory agreement (*e.g.* EDTA, DTPA, acetic acid), this was not always achieved for other procedures, which demonstrated the inadequacy (or the lack of 'maturity') of the scheme(s) at the time it was tested, *e.g.* ammonium acetate and also weak extractants such as calcium chloride or sodium nitrate, owing to difficulties in applications (the low extractable contents resulting in a wide spread of results).[25] Another route for validating extraction methodologies is to make available certified reference materials that are certified on the basis of the schemes in question, *i.e.* for extractable forms of elements.

The findings, proposed procedures and certified reference materials obtained from this programme (Table 1) are presented in detail in the following chapters of the book.

**Table 1** *CRMs available for extractable amounts of heavy metals following single or sequential extraction procedures*

| BCR material | Type of material | Extractant | Analytes |
|---|---|---|---|
| CRM 483 | Sewage sludge-amended soil | EDTA, acetic acid | Cd, Cr, Cu, Ni, Pb, Zn |
| CRM 484 | Sewage sludge-amended soil | EDTA, acetic acid | Cd, Cr, Cu, Ni, Pb, Zn |
| CRM 600 | Calcareous soil | EDTA, DTPA | Cd, Cr, Ni, Pb, Zn |
| CRM 700 | Organic rich soil | EDTA, acetic acid | Cd, Cr, Cu, Ni, Pb, Zn |
| CRM 601 | Lake sediment | BCR-SES (3 steps) | Cd, Cr, Ni, Pb, Zn |
| CRM 701 | Lake sediment | Modified BCR-SES (3 steps) | Cd, Cr, Cu, Ni, Pb, Zn |
| CRM 684 | Lake sediment | Modified Williams scheme (5 steps) | P |

## 1.4   The Variety of Extraction Methods and the Need for Harmonisation

As stressed above, the variety of existing extraction schemes does not allow data to be compared world-wide, which may create problems for data interpretation by regulatory bodies. Efforts are being made to harmonise the approaches used in environmental management.[14] One aspect is collaborative testing (with possible standardisation), as described above, but the consensus for a range of extraction procedures to be possibly used as a common approach by laboratories working in different areas (*e.g.* soils, sludges, composts) is far from being achieved. Some laboratories recognise the suitability of specific schemes based on scientific arguments but they are pressed by national or regional regulations to use other procedures, which creates major confusion. As an example, various tests used in soil surveys, prospecting, characterisation *etc.* are listed below. They focus *e.g.* on 'pseudo-total' contents to assess the extent of pollution in top soils, 'forms' of metals for the evaluation of their mobility, plant availability *etc.*; the list below has been established as a result of consultations of laboratories from different fields:[15]

- *aqua regia* ('pseudo-total' contents) for risk assessment prior to spreading sludge onto agricultural soils;
- EDTA, DTPA, acetic acid, for studies of trace metal mobility, soil–plant transfers, study of physico-chemical processes;
- weak extractants (*e.g.* calcium chloride, calcium nitrate) for plant uptake studies, soil deficiency assessment and remediation, fertility studies, risk assessment;
- column tests, availability batch tests, for environmental risk assessment;
- sodium nitrate, ammonium acetate, for risk assessment and evaluation of soil multi-functionality;
- ammonium chloride, acid oxalate, for the differentiation of lithogenic and anthropogenic origin of some critical elements in soils.

With respect to sediment analysis, extractable forms are determined to assess the heavy metal mobility. Analyses generally concern the 'active' sediment layer and aim at studying the metal remobilisation and fluxes. The concept of obtaining information on (i) easily mobilisable fractions (*e.g.* water or neutral electrolyte solution with or without complexing abilities), (ii) slowly mobilisable fractions (*e.g.* with EDTA or DTPA) and (iii) immobile fractions (*e.g.* with *aqua regia* or real total, using HF) follows in part the approach developed in waste research (total, available, actual leaching). Among the various sequential extraction schemes used nowadays, the three-step sequential extraction protocol developed by BCR is finding increasing international recognition.

## 1.5   Trends and Conclusions

As recently discussed,[26] the coordination between 'harmonisation' (adoption of a common protocol based on scientific grounds, without official mandate) and 'standardisation' activities is not fully satisfactory. This is illustrated by the BCR work on single and sequential extraction schemes for soil and sediment analysis that led to internationally adopted schemes backed-up with CRMs but which, however, were not considered by ISO or CEN as working possible candidates for international standards. There are certainly efforts to be made to establish standards not only on the basis of regulatory requirements but also on the outcome of the results of international collaborative efforts. Furthermore, there is an obvious need to pursue validation activities of existing schemes, *e.g.* by applying them to a wide variety of soil/sediment matrices. It is timely to avoid the multiplication of protocols and critically evaluate newly developed schemes *versus* existing ones. Questions related to their usefulness *etc.* should be thoroughly investigated. Too many procedures are used as a result of regulations without being strongly supported by practical considerations such as easiness, user-friendliness, scientific soundness, robustness *etc.* One of the main trends is certainly the new approach followed by a group of EU experts who collaborate to harmonise various schemes applied to a variety of matrices (waste, soil, sludge, compost, sediment, construction material *etc.*). Such a network, piloted by the SM&T programme, establishes clearly where collaborations are possible among disciplines for adopting common strategies for environmental risk assessment.[15]

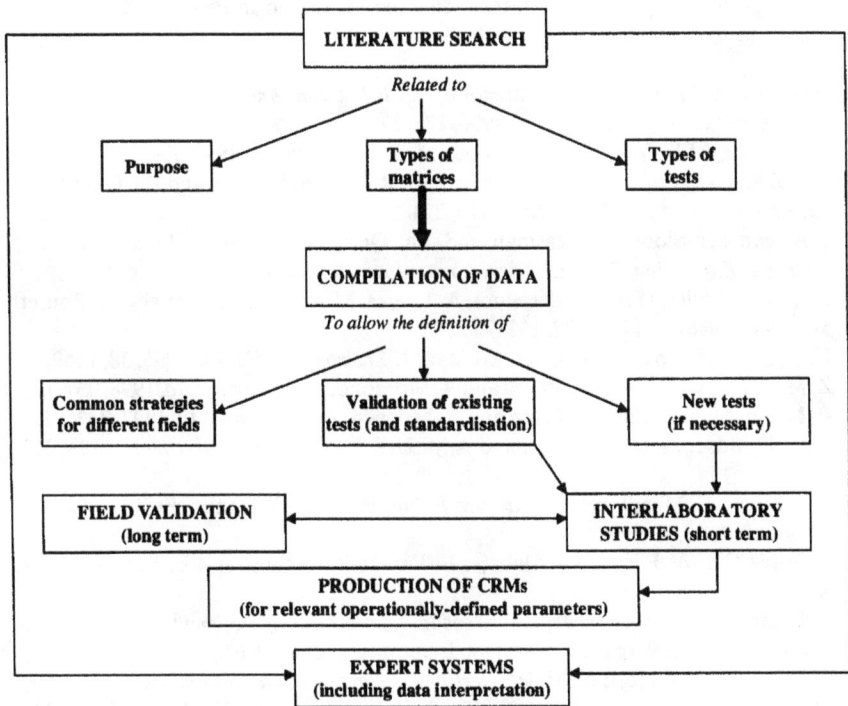

**Figure 2** *Possible database searching system (after Quevauviller[15])*

Standardisation may be the final goal for the establishment of a range of tests 'fit-for-purpose' for various objectives, along the lines of the scheme shown in Figure 2.

Another trend, which is a continuation of the efforts of harmonisation of leaching and extraction tests, is the necessary production of reference materials certified for operationally-defined parameters; this aspect is thoroughly described in this book.

## 1.6 References

1. Ph. Quevauviller, M. Astruc, R. Morabito, F. Ariese and L. Ebdon, *Trends Anal. Chem.*, 2000, **19**, 180.
2. Ph. Quevauviller, M. Filippelli and M. Horvat, *Trends Anal. Chem.*, 2000, **19**, 157.
3. Ph. Quevauviller, A. Ure, H. Muntau and B. Griepink, *Int. J. Environ. Anal. Chem.*, 1993, **51**, 129.
4. Ph. Quevauviller, *Trends Anal. Chem.*, 1998, **17**, 289.
5. D.M. Templeton, F. Ariese, R. Cornelis, L.-G. Danielsson, H. Muntau, H.P. Van Leeuwen and R. Łobiński, *Pure Appl. Chem.*, 2000, **72**, 1453.
6. A. Tessier, P.G.C. Campbell and M. Bisson, *Anal. Chem.*, 1979, **51**, 844.
7. E.A. Thomas, S.N. Luoma, D.J. Cain and C. Johansson, *Water, Air and Soil Pollution*, 1980, **14**, 215.
8. W. Salomons and U. Föstner, *Environ. Technol. Lett.*, 1980, **1**, 506.
9. M. Meguellati, D. Robbe, P. Marchandise and M. Astruc, in *Proc. Int. Conf. Heavy Metals in the Environment*, G. Müller (ed.), CEP Consultants, Heidelberg, 1983, **2**, 1090.
10. A.M. Ure, R. Thomas and D. Littlejohn, *Int. J. Environ. Anal. Chem.*, 1993, **51**, 65.
11. D. McGrath, *Sci. Total. Environ.*, 1996, **178**, 37.
12. S.K. Gupta, M.K. Vollmer and R. Krebs, *Sci. Total Environ.*, 1996, **178**, 11.
13. V. Ruban, J.-F. López-Sánchez, P. Pardo, G. Rauret, H. Muntau and Ph. Quevauviller, *Fresenius J. Anal. Chem.*, 2001, **370**, 224.
14. H.A. van der Sloot, L. Heasman and Ph. Quevauviller (eds.), *Harmonization of Leaching/Extraction Tests*, Elsevier, Amsterdam, 1997, 281.
15. Ph. Quevauviller, H.A. van der Sloot, A. Ure, H. Muntau, A. Gomez and G. Rauret, *Sci. Total Environ.*, 1996, **178**, 133.
16. K. Fytianos, S. Bovolenta and H. Muntau, *J. Environ. Sci. Health*, 1995, **30**, 1169.
17. Z. Mester, C. Cremisini, E. Ghiara and R. Morabito, *Anal. Chim. Acta*, 1998, **359**, 133.
18. B. Marin, M. Valladon, M. Polve and A. Monaco, *Anal. Chim. Acta*, 1997, **342**, 91.
19. C.M. Davidson, P.C.S. Ferreira and A.M. Ure, *Fresenius J. Anal. Chem.*, 1999, **363**, 446.
20. R.A. Nome, C. Mendiguchía-Martínez, F. Nome and H.D. Fiedler, *Environ. Toxicol. Chem.*, 2001, **20**, 693.
21. B. Pérez Cid, A. Fernández Alborés, E. Fernández Gómez and E. Falqué López, *Analyst*, 2001, in press.
22. B. Lombard, in *L'assurance qualité dans les laboratoires agroalimentaires et pharmaceutiques*, M. Feinberg (ed.), Tec&Doc Editions, Paris, 1999, 67.
23. Ph. Quevauviller and E.A. Maier, *Certified Reference Materials and Interlaboratory Studies for Environmental Analysis – The BCR Approach*, Elsevier, Amsterdam, 1999, 558.
24. A. Ure, Ph. Quevauviller, H. Muntau and B. Griepink, *Int. J. Environ. Anal. Chem.*,

1993, **51**, 135.
25. A. Ure, Ph. Quevauviller, H. Muntau and B. Griepink, *EUR Report*, European Commission, Brussels, EN 14472, 1992.
26. Ph. Quevauviller, *J. Soils Sedim.*, 2001, **1**, 175.

CHAPTER 2

# Sequential Extraction Procedures for Sediment Analysis

A. SAHUQUILLO,[1] J.F. LÓPEZ-SÁNCHEZ,[1] G. RAURET,[1]
A.M. URE,[2] H. MUNTAU[3] AND PH. QUEVAUVILLER

[1]University of Barcelona, Department of Analytical Chemistry, Barcelona, Spain
[2]Strathclyde University, Department of Pure and Applied Chemistry, Glasgow, UK
[3]Formerly at the Joint Research Centre, Ispra, Italy

## 2.1 Introduction

Sediments play an important role in aquatic systems both as a sink where contaminants can be stored and as a source of these contaminants to the overlying water and to biota. Due to their ability to sequester metals, sediments are a good indicator of water quality and record the effects of anthropogenic emissions.[1] Thus, sediments are widely used in environmental studies.

Biogeochemical and especially the ecotoxicological significance of a pollutant is determined by its specific binding form and coupled reactivity rather than by its accumulation rate in sediments and due to this fact, the sediment analysis techniques for evaluation of environmental impact of polluted sediments has changed from the determination of total concentrations towards a more sophisticated fractionation of the sediment compounds.[2]

Trace metals in sediments may exist in different chemical forms or ways of binding. In unpolluted soils or sediments trace metals are mainly bound to silicates and primary minerals, relatively immobile species, whereas in polluted ones trace metals are generally more mobile and bound to other sediment phases.[3] However, the measurement or evaluation of this 'speciation, way of binding or fractionation' is very difficult due to the intrinsic complexity of the sediment–water–biota system. Different variables determine the behaviour of these systems: the genesis of the sediment,[4-6] the type of weathering products and the processes which control the transport and redistribution of the elements (adsorption, desorption, precipitation, solubilisation, flocculation, surface complex formation, peptisation, ion exchange, penetration of the crystal structure of minerals, biological mobilisation and immobilisation).[7] Almost all the problems associated with understanding the release processes that control the availability of trace metals concern particle–water interfaces.

There are a variety of well established methods to assess the environmental impact of a given contaminant in sediment–water systems which range from pore water gradient measurements, *in situ* or laboratory incubation experiments up to leaching approaches such as sequential extraction procedures. While the first two approaches aim to study the actual trace metal release potential, the latter approach is widely applied as a predicting tool for the long-term emission potential of sediments.[8]

Different sequential extraction schemes have been proposed for the determination of binding forms of trace metals in sediments. These schemes are a good compromise that provide a practical method for giving information on environmental contamination risks in spite of being operationally defined procedures.

### 2.1.1 Commonly Used Sequential Extraction Procedures for Sediments

The phases considered relevant in heavy metal adsorption in sediments are oxides, sulfides and organic matter. The fractions obtained when applying sequential extraction schemes are related to exchangeable metals, metals mainly bound to carbonates, metals released in reducible conditions such as those bound to hydrous oxides of Fe and Mn, metals bound to oxidisable components such as organic matter and sulfides, and a residual fraction. Table 1 summarises the extractants most commonly used to isolate each fraction.[9–11] Some experimental conditions have been included in the table but there are many variations described involving pH and temperature of the reaction medium, presence of chelating and buffering agents, concentration of the reagent, *etc.*

Although the extractants shown in Table 1 are selective, few of them are specific enough to isolate well-defined fractions and in the different fractions the sediment phases appear overlapped. The fractions can, however, be isolated more specifically by using them in a sequential extraction scheme carried out in a prescribed order.

From the sequential extraction schemes described in the literature, the classical method of Tessier[12–14] has been widely applied in river, marine and stream sediments,[15–16] in municipal composts and sewage sludges,[17–19] and also in soils.[20–22] Several modifications to the Tessier scheme have been proposed by different authors in order to improve the lack of selectivity of the extractant agents towards specific geochemical phases of the sediment. Table 2 shows a scheme of the Tessier sequential extraction procedure together with two modifications made by Förstner[6] towards a more specific isolation of the Fe and Mn oxide and hydroxide phases and by Meguellati[23] who isolates the organically-bound phase before the carbonate-bound.

Other proposed modifications deal with the possibility of accelerating the extraction procedure by the application of ultrasound[24–25] or with the use of

**Table 1** *Commonly used extractants and associated sediment phases*

| Metal fractions | Type of extractant | Extractants used |
|---|---|---|
| Water soluble fraction | Water (distilled or deionised) | Pore water or $H_2O$ extraction |
| Exchangeable and weakly adsorbed fraction | Salts of strong acids and bases or salts of weak acids | $KNO_3$ or $Mg(NO_3)_2$<br>$CaCl_2$ 0.01–0.05 mol $L^{-1}$<br>$MgCl_2$ 1 mol $L^{-1}$ (pH = 7)<br>$BaCl_2$ 1 mol $L^{-1}$ (pH = 7)<br>$NH_4CH_3COO$ 1 mol $L^{-1}$ (pH = 7 or 8.2)<br>$NaCH_3COO$ |
| Carbonate bound fraction | Acid or buffer solutions | $CH_3COOH$ 25% or 1 mol $L^{-1}$<br>$NaCH_3COO$ 1 mol $L^{-1}$ / $CH_3COOH$ (pH = 5)<br>HCl<br>EDTA 0.2 mol $L^{-1}$ (pH = 10–12) |
| Fractions bound to hydrous oxides of Fe and Mn | Reducing solutions<br>Other agents | $NH_4CH_3COO$ 1 mol $L^{-1}$ + 0.2% hydroquinone<br>$NH_2OH.HCl$ 0.02–1 mol $L^{-1}$ in $CH_3COOH$ or $HNO_3$<br>$(NH_4)_2C_2O_8$<br>$(NH_4)_2C_2O_8$ 0.2 mol $L^{-1}$/$H_2C_2O_8$ 0.2 mol $L^{-1}$ in ascorbic acid 0.1 mol $L^{-1}$<br>$Na_2S_2O_4$/Na-citrate/citric acid<br>$Na_2S_2O_4$/Na-citrate/$NaHCO_3$ (DCB)<br>$H_2O_2$ 10% in 0.0001 N $HNO_3$<br>HCl 20%<br>EDTA 0.02–0.1 mol $L^{-1}$ (pH = 8 to 10.5)<br>Hydrazine chloride (pH = 4.5) |
| Organically bound and sulfidic phase | Oxidising reagents | $H_2O_2$ in $HNO_3$ + extraction with $NH_4CH_3COO$ or $MgCl_2$<br>NaClO (pH = 9.5)<br>Alkali pyrophosphate ($Na_4P_2O_7$ or $K_4P_2O_7$)<br>$H_2O_2$/ascorbic acid<br>$HNO_3$/tartaric acid<br>$KClO_3$/HCl |
| Residual fraction | Strong acids | Alkaline fusion<br>$HF/HClO_4/HNO_3$<br>*Aqua regia*<br>$HNO_3/H_2O_2$<br>$HCl/HF/HNO_3$ |

**Table 2** *Tessier's scheme and proposed modifications*

| Sequential extraction procedure | Sediment phases | | | | | |
|---|---|---|---|---|---|---|
| **Tessier** | Exchangeable | Carbonatic | Oxides Fe/Mn | | Organic matter and sulfidic | Residual |
| | Step 1 $MgCl_2$ 1 mol L$^{-1}$ pH = 7 | Step 2 NaOAc 1 mol L$^{-1}$/ HOAc pH = 5 | Step 3 $NH_2OH.HCl$ 0.04 mol L$^{-1}$ in 25% HOAc | | Step 4 $H_2O_2$ 8.8 mol L$^{-1}$/ $HNO_3$ and $NH_4OAc$ 0.8 mol L$^{-1}$ | Step 5 $HF/HClO_4$ |
| **Förstner** | Exchangeable | Carbonatic | Easily reducible | Moderated reducible | Organic matter and sulfidic | Residual |
| | Step 1 $NH_4OAc$ 1 mol L$^{-1}$ pH = 7 | Step 2 NaOAc 1 mol L$^{-1}$/ HOAc pH = 5 | Step 3 $NH_2OH.HCl$ 0.1 mol L$^{-1}$ | Step 4 0.1 mol L$^{-1}$ oxalate buffer | Step 5 $H_2O_2$ 8.8 mol L$^{-1}$ $HNO_3$ and $NH_4OAc$ 0.8 mol L$^{-1}$ | Step 6 $HNO_3$ |
| **Meguellati** | Exchangeable | Organic matter and sulfidic | Carbonatic | Oxides Fe/Mn | | Residual |
| | Step 1 $BaCl_2$ 1 mol L$^{-1}$ pH = 7 | Step 2 $H_2O_2$ 8.8 mol L$^{-1}$ ($HNO_3$) and $NH_4OAc$ 0.8 mol L$^{-1}$ | Step 3 NaOAc 1 mol L$^{-1}$/ HOAc pH = 5 | Step 4 $NH_2OH.HCl$ 0.04 mol L$^{-1}$ in 25% HOAc | | Step 5 Ashing and HF/HCl |

different extractant agents to avoid the readsorption problems observed in oxic sediments.[26]

In addition to these elaborate sequential extraction schemes there are a number of simpler procedures with a smaller number of steps[27-30] which are summarised in Table 3. While the more elaborate schemes are more useful and informative in terms of the physical chemistry of systems and for an understanding of the mechanisms of immobilisation, release and transport, the simpler schemes may well be quite sufficient for practical assessment of the extent of pollution of a sediment and the potential dangers of its use in agriculture, landfill, *etc.* Moreover, optimisation studies are recommended to increase the repeatability when using the longer schemes.[31]

The results coming from the sequential extraction schemes are operationally defined and different mineralogical compositions and organic matter content lead to different efficiencies of extraction and readsorption. An individual method validation is required for each matrix[32] when the aim of the study is to completely isolate sediment fractions.

**Table 3** *Short sequential extraction schemes*

| Sequential extraction procedure | Sediment phases | | | |
|---|---|---|---|---|
| Förstner [27] | Exchangeable and carbonatic | | Organic matter and sulfidic | Residual |
| | Step 1 | | Step 2 | Step 3 |
| | $NH_2OH.HCl$ | | $H_2O_2$ | $HF/HClO_4$ |
| | pH = 2 | | 8.8 mol $L^{-1}$/ | |
| | | | HCl and $NH_4OAc$ | |
| | | | 0.8 mol $L^{-1}$ | |
| Banfi [29] | Exchangeable | Oxides Fe/Mn | Organic matter and sulfidic | Residual |
| | Step 1 | Step 2 | Step 3 | Step 4 |
| | NaOAc | $(NH_4)_2C_2O_8$ | $H_2O_2$ | *Aqua regia* |
| | 1 mol $L^{-1}$/ | 0.2 mol $L^{-1}$/ | 8.8 mol $L^{-1}$/ | |
| | HOAc | $H_2C_2O_8$ | $HNO_3$ and | |
| | pH = 5 | 0.2 mol $L^{-1}$ | $NH_4OAc$ | |
| | | pH = 2 | 0.8 mol $L^{-1}$ | |
| Cosma [30] | Exchangeable | Oxides Fe/Mn | Organic matter and sulfidic | |
| | Step 1 | Step 2 | Step 3 | |
| | $NH_4OAc$ 1 mol $L^{-1}$ | $NH_2OH.HCl$ | $HNO_3$ | |
| | | 1 mol $L^{-1}$ | | |
| | | in 25% HOAc | | |

## 2.2  Preliminary Studies within BCR

The main problem of sequential extraction appears when the results from different procedures are compared because the data, as stated before, depend on experimental conditions.[33] Moreover, the lack of suitable reference materials for this type of analysis does not enable the quality of measurements to be controlled. Results are useful and usable only if they correspond to well-defined and accepted procedures, and with assessed accuracy to make the results comparable elsewhere.[34–35]

The BCR Programme (see Chapter 1) started in 1987 a series of investigations and collaborative studies with the aim of harmonising and improving the methodology for sequential extraction determination of trace metals in sediments.

The information obtained with an initial study of the literature on the topic[36] was discussed by a group of representatives of leading European soil and sediment laboratories and as a conclusion it was considered that the organisation of an interlaboratory trial for testing different sequential extraction schemes was essential as a first step towards the harmonisation of schemes and the preparation of certified reference materials for Cd, Cr, Cu, Ni, Pb and Zn.[37]

Different sequential extraction schemes were tested by four laboratories on seven sediments. The procedures tested were the modified Tessier procedure by Förstner, the Meguellati scheme with six and five steps, respectively (see Table 2), and the short method proposed by Förstner and Salomons with three steps (see Table 3).

The results obtained showed that most of the metals were extracted in the first step in the most mobile phases and the agreement in the obtained information of the schemes was better for highly polluted sediments with extractable metal amounts high enough to avoid analytical problems in the determination.

It was shown that all the tested procedures would be able to classify the predominant nature of the sediments and allow recommendations to be made for their use. Thus, it would be possible to define a simple sequential extraction scheme for the characterisation of the sediments that was practical and selective enough for managerial decision on their use and it was recommended that a reference material be prepared, according to the defined scheme, for quality control purposes.

One aspect to be checked when considering the possibility of preparing reference materials for extractable metal contents is the stability of the solid. A study carried out by Salomons et al.[38] using a sequential extraction scheme with five steps showed that the conclusions drawn on five sediments after an interval of 12 years were similar to management decisions on their use, showing the stability of the extracted fractions.

## 2.3   Proposal for a Common Three-step Sequential Extraction Scheme

The first step for the adoption of a common sequential extraction scheme recommended in the previous project was the organisation of a workshop. The aim of the workshop was to establish the state-of-the-art of extractable trace metal determination, to investigate where limitations existed and discuss the work necessary to overcome these as well as the identified sources of error. The workshop was held in Sitges (Spain) in 1992, in the framework of the project ETMESS (Extractable Trace Metal contents in Soil and Sediment) and a three-step sequential extraction procedure was proposed by a group of European experts.

The scheme was tested in two round-robin exercises with the aim of validating the procedure and demonstrating the feasibility of the preparation of materials that fulfilled the basic requirements of a CRM for stability and homogeneity not as regards the total metal content but for the sequentially extractable metal content.

Two river sediments were selected for the intercomparison exercises with 20 laboratories. The characteristics of the selected materials are shown in Table 4. The materials were prepared according the flow chart shown in Figure 1 by the Environment Institute of the Joint Research Centre in Ispra, Italy. This centre has a wide experience of CRM preparation for environmental purposes in the last 30 years.[39–41]

Many systematic errors were found to be related to calibration errors in the first intercomparison. The reproducibility obtained in the extracts in terms of CVs was quite poor, specially for Ni and Pb. The effects of type of shaking, speed and temperature were checked in the second exercise. It showed an improvement in comparison with the results of the first exercise with CV less than 20% for Cd and Ni in all steps. The discussion of the obtained results in a technical meeting allowed operational details to be added to the protocol that were considered useful from a practical point of view in order to improve the reproducibility among the laboratories. These operational details were: the speed of shaking; special care when adding the oxidant agents in sediments with high organic matter content and; recommendations for the measuring step.[42,43] The scheme used in the interlaboratory trials is shown in Figure 2a.

The results of the homogeneity and stability studies carried out on the sediment used in the second exercise showed moreover that it would be possible to prepare a stable sediment candidate reference material to be certified for the extractable contents by applying the common sequential extraction procedure proposed by the BCR,[44] and thus, the first certification campaign was undertaken. The sediment candidate was a lake sediment collected from different sampling sites of Lake Maggiore (Italy) (see Table 4) and it was prepared according to the same scheme used for preparing the material previously used (Figure 1).

The homogeneity and stability studies carried out with this new sediment material corroborated the feasibility studies about the possibility of certification

**Table 4** *Reference and certified reference materials using the European three-step sequential extraction scheme*

| Sediment sample | Origin | Characteristics | Sequential extraction scheme applied | Sieving |
|---|---|---|---|---|
| 1st interlaboratory trial (River sediment) | Yrseke (The Netherlands) | Siliceous | Original BCR | < 90 μm |
| 2nd interlaboratory trial (River sediment) | River Besòs (Barcelona, Spain) | Calcareous | Original BCR | < 63 μm |
| BCR CRM 601 Lake sediment | Lake Maggiore (Varese, Italy) | Siliceous | Original BCR (certified values) Modified BCR (informative values) | < 90 μm |
| BCR CRM 701 Lake sediment | Lake Orta (Piemonte, Italy) | Siliceous | Modified BCR (certified values) | < 90 μm |

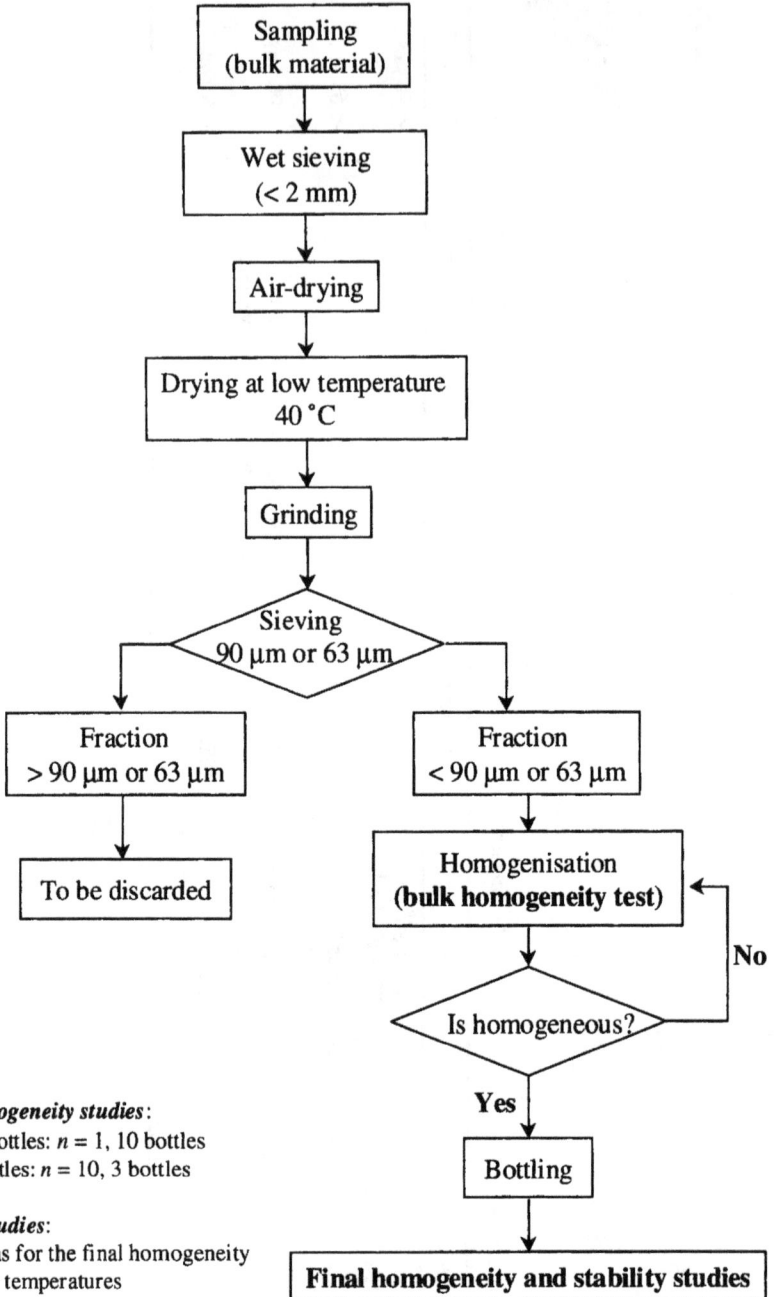

**Figure 1**   *Flow chart for the preparation of reference and certified reference materials*

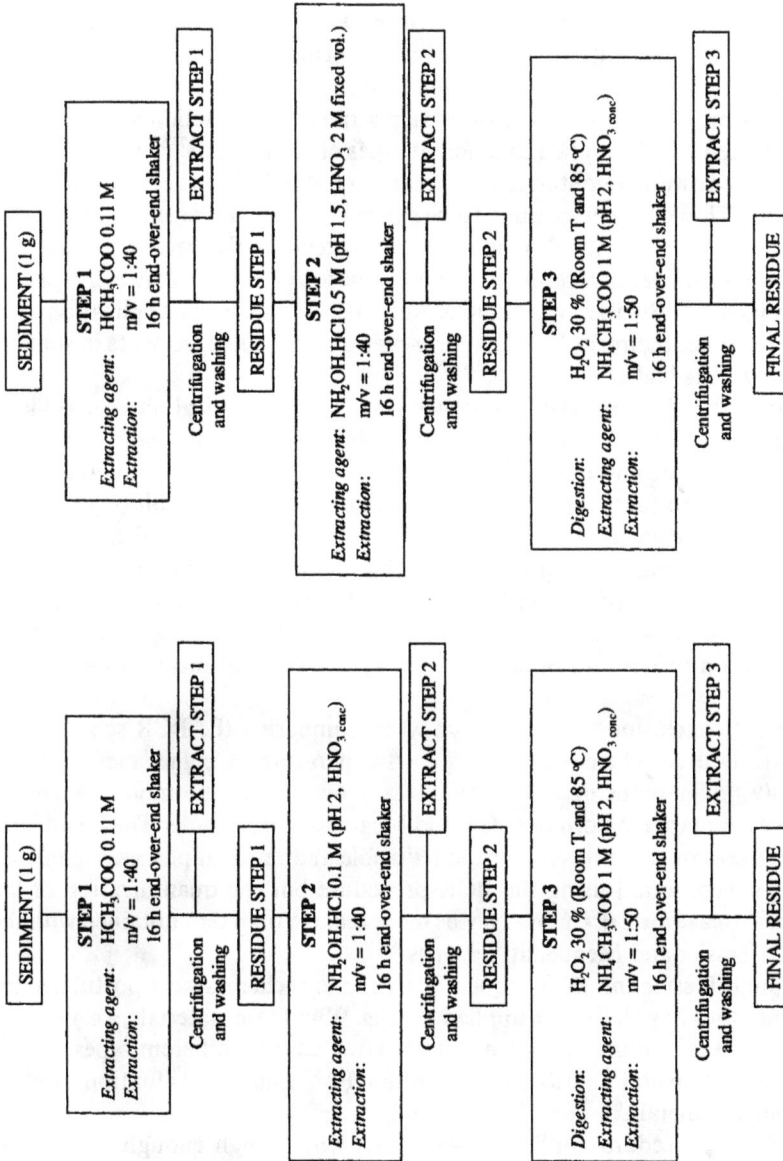

**Figure 2.a**   *Original BCR sequential extraction scheme*

**Figure 2.b**   *Modified BCR sequential extraction scheme*

of extractable trace metal amounts following a standardised sequential three-step extraction procedure.[45] Extractable contents of Cd, Cr, Ni, Pb and Zn were certified in the first step of the procedure, Cd, Ni and Zn in the second step and Cd, Ni and Pb in the third step.[46] Cu in the first step and Pb in the second step were given as indicative values. The rest of the metals were not certified due to the wide spread of the obtained results or owing to a suspicion of instability. The BCR CRM 601 constituted the first available tool for the validation of methodology in the sequential extraction research field.

Since its proposal in 1993 the three-step sequential extraction scheme of the BCR has been widely applied to different types of sediments,[47–54] contaminated soils,[55–57] industrially-contaminated made-up ground,[58] sewage sludges[59] and fly ashes.[60] The scheme was found to be sufficiently repeatable and reproducible for fresh water sediments although it was pointed out the necessity of further investigations to identify the factors responsible for variability between replicate measurements.[47] Smaller variabilities were also found in its application to a marine reference sediment but long-term precision for some elements in some of the extracts was higher than 15%.[50]

As far as the phase selectivity is concerned, the geochemical phase specificity has been shown to be of varying quality as determined upon single substrates. Calcium carbonate and manganese dioxide released most bound metal into the expected reagents (acetic acid and hydroxylammonium hydrochloride, respectively), while organic matter as humic acid generally released metal earlier in the procedure than might be expected.[61,62] The operational nature of the scheme has again been pointed out and several repetitive extractions (up to a total of eight) must be carried out in Steps two and three to completely release the metal bound to iron oxides and organic matter, respectively, in metal-polluted sediments.[63]

Different conclusions can be drawn when comparing the BCR scheme with other widely used schemes according to the literature. A higher metal release, especially under reducing conditions, was shown using the modified Tessier scheme if compared to both the BCR and Meguellati procedures. These two later presented comparative results for the reducible and residual phases. Significant Hg losses were found using the BCR procedure but the quantification of the acetic acid phase for Cd, Cr and Ni was more reliable than that obtained with the modified Tessier and Meguellati schemes.[64]

Although Tessier and BCR sequential extraction schemes were postulated for sediments, they have been also applied to soils. With these materials, a significant correlation was obtained between metals extracted from different types of soils regarding both metal distribution except for Pb[57] and the evaluation of plant availability of metals.[65]

The BCR procedure applied to sediments with a high enough total metal concentration was strongly recommended instead of longer and more difficult procedures for quantifying the fraction of metal characterised by the highest mobility and availability.[51]

In an attempt to simplify the methodology, a recent study compared the results obtained by two sequential extraction procedures (Tessier and BCR schemes) with those estimated from single extractions using identical operating conditions applied in each individual fraction. The test was performed in sewage sludges for Cu, Cr, Ni, Pb and Zn extractable contents. Although for some metals the information obtained was basically the same, the use of single extractions might be only useful for a fast screening of the possible mobility and bioavailability of metals in the environment.[66]

Thus, although some amendments could be proposed to the BCR scheme in order to improve both reproducibility and phase selectivity, the European scheme is a valuable tool for prediction of potential remobilisation of metals and it has proved to be a good compromise between the information obtained and the practicality in its application in the laboratory.

## 2.4 Improvement of the BCR Sequential Extraction Scheme

The difficulties reported when using the BCR sequential extraction scheme relating to the irreproducibility shown in the first certification campaign highlighted the need for refinement of the procedure. Thus, the project 'Trace Metal Extraction from Sediments and Soils (TRAMES)' was undertaken in the framework of the SM&T programme (follow-up of the BCR programme).

A systematic study to assess the sources of uncertainty was carried out focusing mainly on the second step of the sequential extraction procedure and using the lake sediment BCR CRM 601 as the test sample.[67]

The variables and ranges tested are shown in Table 5. Of the variables considered, the pH of $NH_2OH.HCl$ in the second step proved to be the most relevant, especially for Cr, Cu and Pb extraction which showed a dramatic decrease in both extractability and reproducibility as pH increased. The rest of the studied factors did not show significant effect upon reproducibility. From this systematic study the proposed modifications to the scheme were: the use of $NH_2OH.HCl$ 0.5 mol $L^{-1}$ adjusted to pH 1.5 by the addition of a fixed volume of dilute $HNO_3$ to the extractant solution and for all steps the speed of centrifugation was increased from 1500 to 3000 g.

The combination, in the modified scheme, of an increase in the concentration of $NH_2OH.HCl$ from 0.1 to 0.5 mol $L^{-1}$ with a pH lowered to 1.5 also provided a better release of metals bound to hydrous oxides of iron whereas the original reagent largely attacked only the hydrous oxides of manganese.

A small-scale interlaboratory study with eight participating laboratories was undertaken with the aim of comparing the original and the modified protocols using the BCR CRM 601.[68] As a significant improvement concerning the between-laboratory reproducibility was observed for all metals in Step 2 and similar or smaller uncertainties were obtained for Steps 1 and 3, it was agreed to adopt the modified protocol shown in Figure 2b for the preparation of a new certified reference sediment.

The data from this interlaboratory exercise confirmed the stability of extractable contents of Cd, Cr, Ni, Pb and Zn in the existing CRM[69] according to the

**Table 5** *Variables studied as sources of uncertainty in the BCR procedure*

| Variable | Studied range or parameter | Investigated step |
|---|---|---|
| pH of extracting agent | 1.0–3.0 | Step 2 |
| Type of acid used for adjustment | HCl or HNO$_3$ | Step 2 |
| Extraction temperature | 20, 26, 40 °C | Steps 2 and 3 |
| Extraction time | 2–24 h | Step 2 |
| Inert atmosphere | N$_2$ | Step 2 |
| Liquid/solid phase separation | Filtration | Step 2 and 3 |
| | Speed of centrifugation | |
| | Time of centrifugation | |
| | MgCl$_2$ as washing solution | |
| Extractant concentration | 0.1–1 mol L$^{-1}$ | Step 2 |
| Alternative reagents | NH$_4$HC$_2$O$_8$ | Step 2 |
| | H$_2$C$_2$O$_8$ | |

original BCR sequential extraction scheme and gave informative values for the extractable amounts following the modified BCR scheme.

### 2.4.1 New Sediment Certified Reference Material for Extractable Metal Contents

A candidate sediment reference material to be certified according to the modified BCR-scheme was collected from Lake Orta (Piemonte, Italy) and it was prepared in the Environment Institute of the Joint Research Centre in Ispra (Italy) following the validated procedure used for the preparation of CRM 601 (Figure 1). Table 4 shows some characteristics of the sample. The material was certified (BCR CRM 701) in an intercertification campaign for extractable contents of Cd, Cu, Cr, Ni, Pb and Zn in the three steps and it has been available for purchase from the Institute for Reference Materials and Measurements (IRMM) since the beginning of 2001.[70] The studies of homogeneity and stability of the material as well as the certification campaign have been published elsewhere.[71] Figure 3 shows, as an example, the results obtained in the stability studies at 20 °C for the extracted metal in the three steps of the sequential extraction scheme, the residue and the pseudo-total metal content by *aqua regia* extraction. The results for 1, 3, 6 and 12 months show the stability of the material for all metals and fractions.

### 2.4.2 Recent Applications of the Modified BCR Sequential Extraction Scheme

The amended protocol has been recently applied to Cd, Cu, Cr, Ni, Pb and Zn determination in stream sediments from Lake Flumendosa (Italy).[72] The basin of Lake Flumendosa has been mined extensively leaving behind large dumps of mining wastes. Residual metals are mobilised from these wastes and accumulated in the sediments of the lake, known to show the highest Cd concentrations

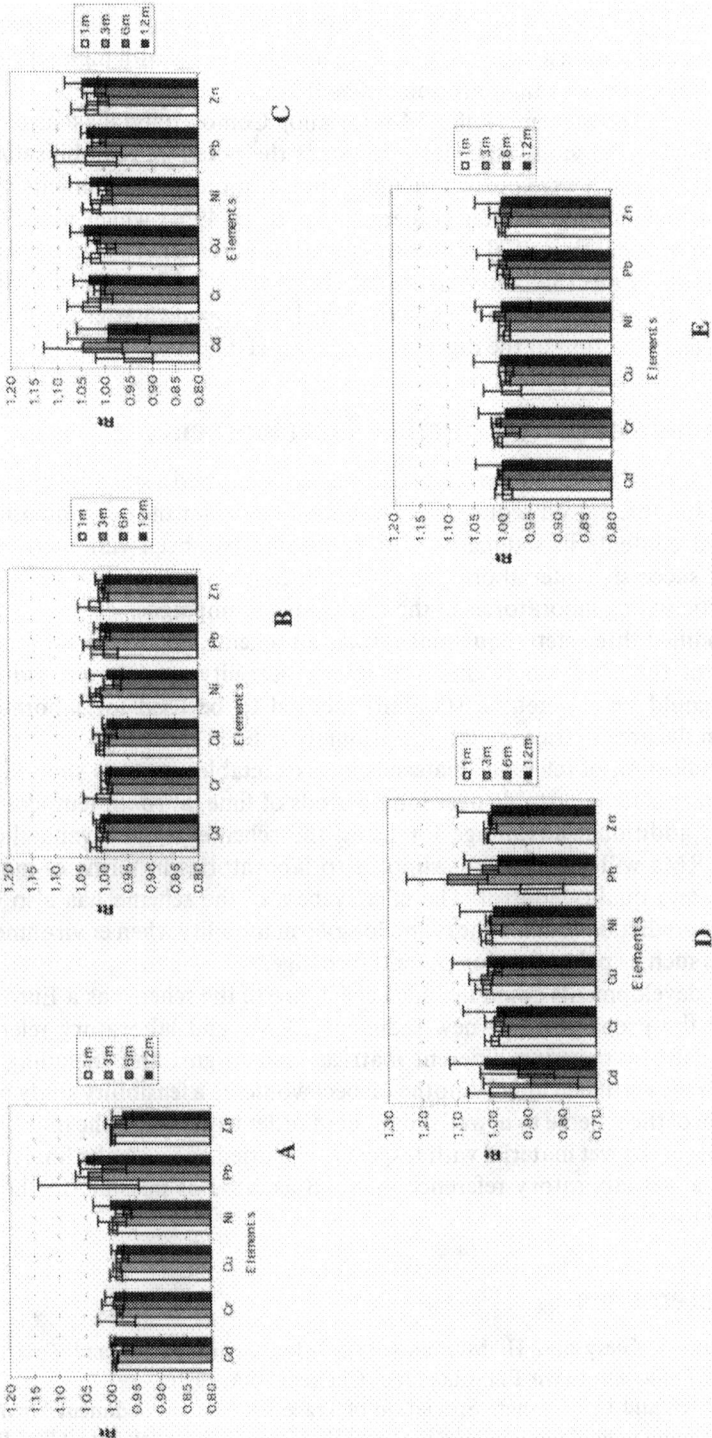

**Figure 3**   *Results from stability studies for BCR 701 at 20°C for: (a) Step 1, (b) Step 2, (c) Step 3, (d) Residue and (e) pseudo-total metal content*

ever encountered in European sediments.[73] The conclusions of this study showed that this procedure can be applied to field samples from different origin, composition and within a wide range of total metal contents with high reproducibility in all steps (most values around 5–6% RSD for triplicates).

Another lake ecosystem (Lake Montorfano, Como, Italy) has also been studied applying the modified scheme to assess the potential remobilisation of Cd, Pb and Zn which were present in high concentrations. The sediments of this lake had a high organic matter content (from 16 to 49%) which made them extremely reactive at the initial addition of the oxidant reagent used in Step 3 and could present serious difficulties as stated elsewhere.[48] The modified protocol has proved to be reproducible even with these high organic matter contents by paying special attention to the digestion step with $H_2O_2$.[74]

## 2.5 Conclusions and Further Developments

The framework of the BCR/SM&T programme has allowed a stepwise approach to select a suitable agreed sequential extraction scheme for the determination of trace metal contents in sediments. The procedure has been validated in the context of successive interlaboratory studies in terms of agreement among the 15–20 participating laboratories in the certification campaigns.

The modified three-step sequential extraction scheme, due to the high reproducibility of the obtained results and its applicability to different sediment matrices, could be a potential standard method to be used for laboratories working in the area of trace metal operationally-defined speciation.

The certification of reference materials for extractable amounts showed that dried sediments remain stable over long periods of time (at room temperature). This is an additional advantage for using the scheme as the availability of suitable CRMs will offer a great support to laboratories in terms of method validation and quality control. The effectiveness of the scheme has also been shown as a prediction tool for short and long-term mobility when environmental conditions such as pH and redox potential change.

Further developments could include acceptance of the scheme as a European norm and the preparation of new sediment CRMs and laboratory reference materials (LRMs) covering different matrices and origins and including new elements in the certified values. Another aspect would be a feasibility study of the application of the scheme to a 'wet' sediment in order to ascertain the mobility of trace metals in the wet material with respect to the dried one with the final aim of preparing a wet laboratory reference material as close as possible to the real samples in this field.

## 2.6 References

1. R. Baudo, J. Giesy and H. Muntau (eds.), *Sediments: Chemistry and Toxicity of In-place Pollutants*, Lewis Publishers Inc., Chelsea, USA, 1990.
2. M. Kersten and U. Förstner, 'Speciation of Trace Elements in Sediments' in *Trace Element Speciation: Analytical Methods and Problems*, G.E. Batley (ed.), CRC Press,

Boca Raton, Florida, 1989.
3. G. Rauret, *Talanta*, 1998, **46**, 449.
4. H.J.M. Bowen, *Trace Elements in Biochemistry*, Academic Press, London, 1966.
5. H.J.M. Bowen, *Environmental Chemistry of the Elements*, Academic Press, London, 1979.
6. U. Förstner, *Intern. J. Environ. Anal. Chem.*, 1993, **51**, 5.
7. J.A. Plant and R. Raiswell, 'Principles of Environmental Geochemistry' in *Applied Environmental Geochemistry*, I. Thornton (ed.), Academic Press, London, 1983.
8. H.A. van der Sloot, L. Heasman and Ph. Quevauviller (eds.), *Harmonization of Leaching/Extraction Tests*, Elsevier Science B.V., Amsterdam, 1997.
9. A. Ure, Ph. Quevauviller, H. Muntau and B. Griepink, Report EUR 14763 EN, 1993.
10. T.T. Chao, *J. Geochem. Explor.*, 1984, **20**, 101.
11. A.K. Das, R. Chakraborty, M.L. Cervera and M. de la Guardia, *Talanta*, 1995, **42**, 1007.
12. A. Tessier, P.G.C. Campbell and M. Bisson, *Anal. Chem.*, 1979, **51**, 844.
13. A. Tessier, P.G.C. Campbell and M. Bisson, *J. Geochem. Explor.*, 1982, **16**, 77.
14. A. Tessier, F. Rapin and R. Carignan, *Geochimica et Cosmochimica Acta*, 1985, **49**, 183.
15. J. Zhang, W.W. Huang and Q. Wang, *Chemical Geology*, 1994, **112**, 275.
16. J.F. López-Sánchez, R. Rubio, C. Samitier and G. Rauret, *Wat. Res.*, 1996, **30**, 153.
17. J. Chwastowska and K. Skalmowski, *Intern. J. Environ. Anal. Chem.*, 1997, **68**, 13.
18. R. Zufiaurre, A. Olivar, P. Chamorro, C. Nerín and A. Callizo, *Analyst*, 1998, **123**, 255.
19. I. Lavilla, B. Pérez-Cid and C. Bendicho, *Anal. Chim. Acta*, 1999, **381**, 297.
20. X. Li, B.J. Coles, M.H. Ramsey and I. Thornton, *Chemical Geology*, 1995, **124**, 109.
21. X. Li, B.J. Coles, M.H. Ramsey and I. Thornton, *Analyst*, 1995, **120**, 1415.
22. I. Maiz, I. Arambarri, R. Garcia and E. Millán, *Environmental Pollution*, 2000, **110**, 3.
23. M. Meguellati, D. Robbe, P. Marchandise and M. Astruc, *Proc. Int. Conf. Heavy Metals in the Environment*, Heidelberg, CEP Consultants, Edinburgh, 1983, **2**, 1090.
24. B. Pérez-Cid, I. Lavilla and C. Bendicho, *Fresenius J. Anal. Chem.*, 1999, **363**, 667.
25. B. Pérez-Cid, I. Lavilla and C. Bendicho, *Intern. J. Environ. Anal. Chem.*, 1999, **73**, 79.
26. J.L. Gómez-Ariza, I. Giráldez, D. Sánchez-Rodas and E. Morales, *Anal. Chim. Acta*, 1999, **399**, 295.
27. U. Förstner and W. Salomons, *Environ. Technol. Lett.*, 1980, **1**, 494.
28. W. Salomons and U. Förstner, *Metals in the Hydrocycle*, Springer-Verlag, Berlin, 1984.
29. C. Banfi, R. Cenci, M. Bianchi and H. Muntau, Istituto dell'Ambiente, Centro Comune di Ricerca, Ispra, EUR Report, 1992.
30. B. Cosma, R. Frache, A. Mazzucotelli and F. Soggia, *Annali di Chimica*, 1991, **81**, 371.
31. G.M. Accomasso, V. Zelano, P.G. Daniele, D. Gastaldi, M. Ginepro and G. Ostacoli, *Spectrochim. Acta*, 1993, **49A**, 1205.
32. P.M.V. Nirel and F.M.M. Morel, *Wat. Res.*, 1990, **24**, 1055.
33. J.F. López-Sánchez, R. Rubio and G. Rauret, *Intern. J. Environ. Anal. Chem.*, 1993, **51**, 113.
34. B. Griepink, *Intern. J. Environ. Anal. Chem.*, 1993, **51**, 123.
35. Ph. Quevauviller, *Trends in Anal. Chem.*, 1998, **17**, 632.
36. A. Ure, Ph. Quevauviller, H. Muntau and B. Griepink, EUR Report 14763 EN, European Commission, Brussels, 1992.
37. A. Ure, Ph. Quevauviller, H. Muntau and B. Griepink, *Intern. J. Environ. Anal. Chem.*, 1993, **51**, 135.
38. W. Salomons and S. Scheltens, Report T44, Inst. for Soil Fertility, Haren, 1987.

39. G.N. Kramer, H. Muntau, E. Maier and J. Pauwels, *Fresenius J. Anal. Chem.*, 1998, **360**, 299.

40. B. Griepink, E.A. Maier, Ph. Quevauviller and H. Muntau, *Fresenius J. Anal. Chem.*, 1991, **339**, 599.

41. *BCR Catalogue of Reference Materials*, European Commission, Joint Research Centre, Institute for Reference Materials and Measurements (IRMM), 2000.

42. H. Fiedler, PhD Thesis, Departament de Química Analítica, Universitat de Barcelona, 1995.

43. Ph. Quevauviller, G. Rauret, H. Muntau, A.M. Ure, R. Rubio, J.F. López-Sánchez, H.D. Fiedler and B. Griepink, *Fresenius J. Anal. Chem.*, 1994, **349**, 808.

44. H. Fiedler, J.F. López-Sánchez, R. Rubio, G. Rauret, Ph. Quevauviller, A.M. Ure and H. Muntau, *Analyst*, 1994, **119**, 1109.

45. Ph. Quevauviller, G. Rauret, J.F. López-Sánchez, R. Rubio, A.M. Ure and H. Muntau, *Sci. Tot. Environ.*, 1997, **205**, 223.

46. Ph. Quevauviller, G. Rauret, J.F. López-Sánchez, R. Rubio, A.M. Ure and H. Muntau, EUR Report 17554 EN, BCR CRM 601 certification report, European Commission, 1997.

47. C.M. Davidson, R.P. Thomas, S.E. McVey, R. Perala, D. Littlejohn and A.M. Ure, *Anal. Chim. Acta*, 1994, **291**, 277.

48. A. Sahuquillo, J.F. López-Sánchez, R. Rubio, G. Rauret and V. Hatje, *Fresenius J. Anal. Chem.*, 1995, **351**, 197.

49. A.U. Belazi, C.M. Davidson, G.E. Keating, D. Littlejohn and M. McCartney, *J. Anal. Atomic Spectrom.*, 1995, **10**, 233.

50. B. Marin, M. Valladon, M. Polve and A. Monaco, *Anal. Chim. Acta*, 1997, **342**, 91.

51. Z. Mester, C. Cremisini, E. Ghiara and R. Morabito, *Anal. Chim. Acta*, 1998, **359**, 133.

52. R. Martin, D.M. Sánchez and A.M. Gutiérrez, *Talanta*, 1998, **46**, 1115.

53. J. Userno, M. Gamero, J Morillo and I. Gracia, *Environ. Int.*, 1998, **24**, 487.

54. S. Tokalioglu, S. Kartal and L. Elçi, *Anal. Chim. Acta*, 2000, **413**, 33.

55. M. Raksasataya, A.G. Langdon and N.D. Kim, *Anal. Chim. Acta*, 1996, **332**, 1.

56. I. Maiz, M.V. Esnaola and E. Millan, *Sci. Tot. Environ.*, 1997, **206**, 107.

57. M.D. Ho, G.J. Evans, *Anal. Comm.*, 1997, **34**, 363.

58. C.M. Davidson, A.L. Duncan, D. Littlejohn, A.M. Ure and L.M. Garden, *Anal. Chim. Acta*, 1998, **363**, 45.

59. B. Perez-Cid, I. Lavilla and C. Bendicho, *Analyst*, 1996, **121**, 1479.

60. M.D. Petit and M.I. Rucandio, *Anal. Chim. Acta*, 1999, **401**, 283.

61. C. Whalley and A. Grant, *Anal. Chim. Acta*, 1994, **291**, 287.

62. P.P. Cooetze, K. Gouws, S. Pluddemann, M. Yacoby, S. Howell and L. Dendrijver, *Water SA*, 1995, **21**, 51.

63. J.L. Gómez-Ariza, I. Giráldez, D. Sánchez-Rodas and E. Morales, *Anal. Chim. Acta*, 2000, **414**, 151.

64. J.L. Gómez-Ariza, I. Giráldez, D. Sánchez-Rodas and E. Morales, *Sci. Tot. Environ.*, 2000, **246**, 271.

65. T. Zhang, X. Shan and F. Li, *Commun. Soil Sci. Plant Anal.*, 1998, **29**, 1023.

66. A. Fernández Alborés, B. Pérez-Cid, E. Fernández-Gómez and E. Falqué-López, *Analyst*, 2000, **125**, 1353.

67. A. Sahuquillo, J.F. López-Sánchez, R. Rubio, G. Rauret, R.P. Thomas, C.M. Davidson and A.M. Ure, *Anal. Chim. Acta*, 1999, **382**, 317.

68. G. Rauret, J.F. López-Sánchez, A. Sahuquillo, R. Rubio, C. Davidson, A. Ure and Ph. Quevauviller, *J. Environ. Monit.*, 1999, **1**, 57.

69. J.F. López-Sánchez, A. Sahuquillo, H.D. Fiedler, R. Rubio, G. Rauret, H. Muntau

and Ph. Quevauviller, *Analyst*, 1998, **123**, 1675.
70. G. Rauret, J.F. López-Sánchez, D. Lück, M. Yli-Halla, H. Muntau and Ph. Quevauviller, EUR 19775 EN, certification report of BCR 701, European Commission, 2001.
71. M. Pueyo, G. Rauret, D. Lück, M. Yli-Halla, H. Muntau, Ph. Quevauviller and J.F. López-Sánchez, *J. Environ. Monit.*, 2001, **3**, 243.
72. A. Sahuquillo, D. Pinna, G. Rauret and H. Muntau, *Fres. Environ. Bulletin*, 2000, **9**, 360.
73. M. Cireddu, S. Fadda, M. Fiori, S.M. Grillo, M.G. Manca, O. Masala, A. Marcello and S. Pretti, *VIII Congreso Geologico Chileno*, 1997, Actas 1, 675.
74. F. Serano, A. Sahuquillo, M. Bianchi and H. Muntau, European Commission, Joint Research Center, 2000, Euro Report EUR 19625 IT.

CHAPTER 3

# Extraction Procedures for Soil Analysis

J.F. LÓPEZ-SÁNCHEZ, A. SAHUQUILLO, G. RAURET,
M. LACHICA,[1] E. BARAHONA,[1] A. GOMEZ,[2] A.M. URE,
H. MUNTAU AND PH. QUEVAUVILLER

[1]Estación Experimental del Zaidin, Granada, Spain
[2]NRA, Station d'Agronomie, Lermave, Villenave d'Ornon, France

## 3.1 Introduction

Trace elements in soils appear in different chemical forms or ways of binding. In unpolluted soils, trace elements exist mainly as relatively immobile species in silicates, aluminates and other primary minerals but as a result of weathering the trace element content is gradually mobilised to forms available to plants. In polluted soils, metals are mainly in more labile forms (sorbed, complexed, co-precipitated, *etc.*) and contribute to the pool of potentially available metals. The most relevant trace elements are arsenic, boron, cadmium, chromium, cobalt, copper, lead, molybdenum, nickel, selenium, titanium, vanadium and zinc. Some of these trace metals such us chromium, nickel or zinc are essential to plant growth but they have toxic effects at high levels. Others, such cadmium or lead, are non-essential and potentially toxic.

In some kinds of studies (environmental, agricultural, geochemical, *etc.*) the determination of these ways of binding in soils and sediments provides more useful information on element mobility and availability than the determination of the total content. However, the determination of the different ways of binding is quite difficult due to the complexity of the analysed matrix and is often impossible. Different analytical approaches are used: many of them focus on element desorption from the solid phase using chemical reagents; others are focused on the element adsorption from a solution by the solid phase or in the use of instrumental techniques such as X-ray.[1] Among them, the approaches based on extraction/leaching procedures are the most widely accepted and used. During the last decades, extraction procedures for extractable heavy metals in soils have been developed and modified. In this respect, two groups of tests must be considered: the single reagent extraction test – one extraction solution and one soil sample – and the sequential extraction procedures – several extraction solutions are used sequentially on the same sample – although this last type of extraction is still in development for soils. Both types of extractions are applied, using not only different extraction schemes but also different laboratory conditions. This leads to the use of a great deal of extraction procedures.

### 3.1.1 Single Extraction Procedures

Extraction procedures using a single extractant are widely used in soil science. These procedures are designed to dissolve element contents correlated with the availability of the element to the plants. This approach is well established for major nutrients and it is commonly applied in studies of fertility and quality of crops. The approach is also applied to predict the plant uptake of essential elements, to determine element deficiency or excess in a soil, to study the physical-chemical behaviour of elements in soils or in survey purposes. They are also applied, to a lesser extent, to elements considered as pollutants, such as heavy metals. The application of extraction procedures to polluted or naturally contaminated soils is mainly focused to ascertain the potential availability and mobility of the pollutants and their migration in a soil profile, which is usually connected with groundwater problems.[2] As far as soil is concerned, single extraction procedures are always restricted to a reduced group of elements and they are applied to a particular type of soil: siliceous, carbonated or organic. In a regulatory context, two applications for leaching tests can be recognised: the assessment or prediction of the environmental effects of a pollutant concentration in the environment and the promulgation of guidelines or objectives for soil quality as, for example, for land application of sewage sludge or dredged sediments. The data obtained when applying these tests are used for decision-makers in topics such as land use of soil or in countermeasures application.[3] Table 1 shows a summary of the most common leaching tests used in soil analysis.

From the table it can be observed that single extraction includes a large spectrum of extractants. They range from very strong acids, such as *aqua regia*, nitric acid or hydrochloric acid, to neutral unbuffered salt solutions, mainly $CaCl_2$ or $NaNO_3$. Other extractants such as buffered salt solutions or complexing agents, because of their ability to form very stable water-soluble complexes with a wide range of cations, are frequently applied. For boron, hot water is also used. Basic extraction by using sodium hydroxide is used to assess the influence of the dissolved organic carbon in the release of heavy metals from soils. Information and details about a large number of extractants has been reviewed by Pickering[14] and Lebourg.[15]

The increasing performance of the analytical techniques used for element determination in an extract, together with the increasing evidence that exchangeable metals better correlate with plant uptake, has lead extraction methods to evolve towards the use of less and less aggressive solutions.[11] These solutions are sometimes called soft extractants and are based on non-buffered salt solutions,

**Table 1**   *Extraction tests used in soil analysis*

| Group | Type and solution strength | References |
|---|---|---|
| Acid extraction | $HNO_3$ 0.43–2 mol $L^{-1}$ | 4 |
| | *Aqua regia* | 5 |
| | HCl 0.1–1 mol $L^{-1}$ | 4 |
| | $CH_3COOH$ 0.1 mol $L^{-1}$ | 6 |
| | Melich 1: HCl 0.05 mol $L^{-1}$ + | 7 |
| | $H_2SO_4$ 0.0125 mol $L^{-1}$ | |
| Chelating agents | EDTA 0.01–0.05 mol $L^{-1}$ at different pH | 4 |
| | DTPA 0.005 mol $L^{-1}$ + TEA 0.1 mol $L^{-1}$ | 8 |
| | + $CaCl_2$ 0.01 mol $L^{-1}$ | |
| | Melich 3: $CH_3COOH$ 0.02 mol $L^{-1}$ + $NH_4F$ | 9 |
| | 0.015 mol $L^{-1}$ + $HNO_3$ 0.013 mol $L^{-1}$ | |
| | + EDTA 0.001 mol $L^{-1}$ | |
| Buffered salt solution | $NH_4$-acetate, acetic acid buffer pH = 7 | 10 |
| | 1 mol $L^{-1}$ | |
| | $NH_4$-acetate, acetic acid buffer pH = 4.8 | 4 |
| | 1 mol $L^{-1}$ | |
| Unbuffered salt solution | $CaCl_2$ 0.1 mol $L^{-1}$ | 4 |
| | $CaCl_2$ 0.05 mol $L^{-1}$ | 4 |
| | $CaCl_2$ 0.01 mol $L^{-1}$ | 4 |
| | $NaNO_3$ 0.1 mol $L^{-1}$ | 11 |
| | $NH_4NO_3$ 1 mol $L^{-1}$ | 4 |
| | $AlCl_3$ 0.3 mol $L^{-1}$ | 12 |
| | $BaCl_2$ 0.1 mol $L^{-1}$ | 13 |

although diluted acids and complexant agents are also included. Neutral salts dissolve mainly the cation-exchangeable fraction although in some cases the complexing ability of the anion can play a certain role. Diluted acids solubilise partially trace elements associated with different fractions such as exchangeable, carbonates, iron and manganese oxides, and organic matter. Complexing agents solubilise not only the exchangeable element fraction but also the element fraction forming organic matter complexes and the element fraction fixed on the soil hydroxides. Nowadays, it is generally accepted that extractants are not selective and that minor variations in analytical procedures have significant effect on the results. According to Lebourg[15] some of these methods have been adopted officially or the adoption is under study in different countries with different objectives. An account of these methods is given in Table 2.

## 3.1.2   Sequential Extraction Procedures

These procedures are widely applied for sediment analysis (see Chapter 2 of this book) and are focused on differentiating the different association forms of metals in the solid phases. To do so, several extracting reagents are applied sequentially

**Table 2**  *Extraction methods standardised or proposed for standardisation in several European countries*

| Country | Method | Objective | Reference |
|---|---|---|---|
| Germany | 1 mol L$^{-1}$ NH$_4$NO$_3$ | Mobile trace element determination | 16 |
| France | 0.01 mol L$^{-1}$ Na$_2$-EDTA + 1 mol L$^{-1}$ CH$_3$COONH$_4$ at pH = 7 | Available Cu, Zn and Mn evaluation for fertilisation purposes | 17 |
|  | DTPA 0.005 mol L$^{-1}$ + TEA 0.1 mol L$^{-1}$ + CaCl$_2$ 0.01 mol L$^{-1}$ at pH = 7.3 |  |  |
| Italy | 0.02 mol L$^{-1}$ EDTA + 0.5 mol L$^{-1}$ CH$_3$COONH$_4$ at pH = 4.6 | Available Cu, Zn, Fe and Mn evaluation in soils | 18 |
|  | DTPA 0.005 mol L$^{-1}$ + TEA 0.1 mol L$^{-1}$ + CaCl$_2$ 0.01 mol L$^{-1}$ at pH = 7.3 |  |  |
| Netherlands | CaCl$_2$ 0.1 mol L$^{-1}$ | Availability and mobility of heavy metals in polluted soils evaluation | 19 |
| Switzerland | NaNO$_3$ 0.1 mol L$^{-1}$ | Soluble heavy metal (Cu, Zn, Cd, Pb and Ni) determination and ecotoxicity risk evaluation | 20 |
| United Kingdom | EDTA 0.05 mol L$^{-1}$ at pH = 4 | Cu availabitity evaluation | 21 |

to the sample according the following order: unbuffered salts, weak acids or buffered salts, reducing agents, oxidising agents and strong acids.

During the last decade, there was increasing interest on applying sequential extraction to study trace metal partitioning in soils, although first studies on nutrient element fractionation were already reported some decades ago.[22-24] Most of the published literature is based on the work of Tessier,[25] but new approaches, improvements and/or modifications are also proposed,[26-30] *i.e.* the modification of the procedure to allow multielemental determination by ICP-AES[31] or to analyse a calcareous matrix.[32] In this way, the work carried out in Europe to develop a harmonised sequential extraction procedure is remarkable, *i.e.* the so called BCR procedure,[33-35] that is gaining acceptance among the scientists using such procedures.[36-39] There are also studies dealing with problems of sequential extraction schemes already reported for sediments, such as redistribution of metal fractions during the extraction process[40-42] or lack of selectivity when dissolving the different soil phases.[43]

However, most of the work carried out is focused on the use of sequential extraction as a tool to evaluate the availability of metals to plants[44-50] or to study metal distribution and/or mobility in polluted, forest and agricultural soils.[51–57] In the first type of studies, most of the results show that some correlations exist between the exchangeable and the acid-soluble fractions and plant uptake, although the relationship seems to be dependant on the type of soil and the type of plant. These results indicate that sequential extraction procedures, though operationally defined, may provide complementary and valuable data in order to predict metal availability to plants. In relation to the mobility and/or pollution studies, the main conclusion that can be drawn is that sequential extraction is useful to determine contamination problems, because metals from anthropogenic sources are retained by non-residual fractions and, consequently, are more mobile than those from the parent soil materials. On the other hand, Cd, Cu and Zn appear as mobile metals, whereas Cr, Ni and Pb are more strongly retained.

Moreover, there is literature about other uses of sequential extractions in soil science. There are studies on the heavy metal retention by silty soils,[58] the relationships between metal sorption and chemical forms[59] or the evaluation of the efficacy of restoration strategies using sequential extraction.[60,61]

## 3.2 Preliminary Studies

After a study of the literature carried out by A.M. Ure,[62] a consultation with European experts in both soil and sediment disciplines was organised on behalf of BCR to determine whether there was a need for BCR involvement in harmonising the methodology to determine metal partitioning in soils and sediments and, if so, to make recommendations on the way to proceed. The report and its recommendations were discussed at a meeting in Brussels in 1987 and it was considered that the following actions, with respect to soil analysis, could be initiated by BCR:

(a) An interlaboratory comparative analysis of soil using single extraction procedures;

(b) The preparation of a soil material from sludge-contaminated soil in air-dried, <2 mm, ungrounded form;

(c) The organisation of an interlaboratory study using prescribed protocols detailing the procedures;

(d) It was agreed that, provided the methodologies were adequate or could be made so by the interlaboratory collaborative study, there was both a need for, and a sufficient demand from, soil and environmental laboratories for soil reference materials whose contents had been certified by such procedures.

### 3.2.1 Stability of Soil Extractable Contents by Single Extractants

For a soil to be certified for the concentrations of extractable forms, the verification of stability is of paramount importance to ensure that the concentrations measured at regular intervals will remain stable over a period of years. The reliability and repeatability of many of the procedures for the extraction of soils for the diagnosis and prediction of deficiencies of essential nutrient elements such as Cu and Zn in agricultural laboratories were well established. For example, the analysis of eight top soils, dried at 25 °C, for the 0.05 mol L$^{-1}$ EDTA-extractable Cu, repeated some ten times over a period of seven years by different analysts, provided a mean CV of 6.7% with a range from 4.5% to 9.4%.[63,64] Such a degree of stability was to be expected for many of these essential trace element extraction procedures. For heavy metals and other toxic elements, the procedures were less well established and, while there was a general consensus that many of these provide extractable contents that are constant over time, much of the evidence for this was anecdotal. For some elements such as Hg and Se, there is the possibility of losses by volatilisation, while for elements, such as Mn, the variability in the oxidation state may contribute to changes in the extractability. Considerable problems were likely for Mn: although Boken[65] found little change in exchangeable Mn (MnNO$_3$ 1 mol L$^{-1}$ extraction), others, including Fujimoto and Sherman,[66] and Berndt[67] found substantial increases on air-drying. More importantly, considerable increases in extractable (CaCl$_2$ 0.01 mol L$^{-1}$ extraction) Mn occurred in storage at 20 °C, which continued over period of months. While storage at 4 °C improved the situation, even this was not a complete answer. As this, perhaps extreme, example shows, the stability of the extractable content of an element will be influenced by the nature of the sample, whether it is stored in wet or dry conditions, by the temperature of drying and of storage and so on.[68–70]

For these reasons, it was considered necessary to carry out a trial of the stability of a number of single soil extraction procedures for a range of elements. The temporal stability of the extractable contents of soil was investigated using an air-dried, sieved at 2 mm mesh, sewage sludge-amended soil.[33,62] The elements Cd, Cr, Cu, Fe, Mn, Ni, Pb and Zn were determined in EDTA 0.05 mol L$^{-1}$, acetic acid 0.43 mol L$^{-1}$, calcium chloride 0.1 mol L$^{-1}$ and ammonium acetate 1 mol L$^{-1}$ extracts of this soil by AAS over a one-year period (1989–1990). EDTA extracts were also analysed for Cd, Cr, Cu, Mn, Pb and Zn at the time of collection of the soil (1986) so that for this extractant an (almost) three-year comparison was available in addition to the one-year comparisons described above. The results are presented in Table 3 and the percentage changes with time recorded in Table 4.

From Table 4, it can be seen that for EDTA extraction the one-year changes were all less than 9% with the exception of Cr ($-31\%$) and Zn ($+14\%$). These changes in the contents of Cr and Zn did not reflect a continuous trend since the three-year changes were very small indeed, *i.e.* around 3% for both elements. These aberrant values for Cr and Zn were, therefore, probably values arising from imprecision in analysis, sampling errors, contamination or a rapid change in the partitioning upon drying after which the extractable contents became constant. It was concluded that, overall, EDTA extraction provides contents that were stable with time within about 10% for all the elements tested. This conclusion is in agreement with earlier published findings for EDTA-extractable Cu over eight years.[64] Acetic acid extraction was temporarily less stable than EDTA, but the variation is within 20% for all the elements except Cr ($+30\%$)

**Table 3** *Stability test. Comparison of analysis carried out on EDTA 0.05 mol $L^{-1}$, acetic acid 0.43 mol $L^{-1}$, ammonium acetate 1 mol $L^{-1}$ and calcium chloride 0.1 mol $L^{-1}$ extracts of sludge-amended soils in July 1989 and July 1990. A limited comparison is also made with EDTA contents determined in October 1986. Concentrations (mg $kg^{-1}$ air-dry soil) are means of three determinations and CV is in percentage of the mean*

| Element | EDTA | | | HOAc | | CaCl₂ | | NH₄OAc | |
| | 1986 | 1989 | 1990 | 1989 | 1990 | 1989 | 1990 | 1989 | 1990 |
|---|---|---|---|---|---|---|---|---|---|
| Cd (mg kg⁻¹) | 18.3 | 22.6 | 21.8 | 16.3 | 17.5 | 1.00 | 2.2 | 2.3 | 2.4 |
| CV (%) | 1.7 | 1.0 | 2.5 | 1.4 | 0.9 | 0.05 | 3.5 | 2.6 | 2.4 |
| Cr (mg kg⁻¹) | 7.8 | 10.9 | 7.5 | 20.0 | 27.3 | <0.2 | 0.1 | 1.0 | 1.9 |
| CV (%) | 4.5 | 1.6 | 3.1 | 0.0 | 1.6 | – | 0.0 | 6.6 | 10.8 |
| Cu (mg kg⁻¹) | 156 | 152 | 160 | 22.2 | 26.8 | 0.3 | 0.8 | 3.1 | 5.5 |
| CV (%) | 1.5 | 2.1 | 2.1 | 0.5 | 0.9 | 4.9 | 0.0 | 1.8 | 6.4 |
| Fe (mg kg⁻¹) | – | 1813 | 1817 | 49.3 | 40.3 | <0.5 | 0.5 | 5.2 | 12.0 |
| CV (%) | – | 7.5 | 6.1 | 1.2 | 5.7 | – | 6.0 | 38.2 | 41.0 |
| Mn (mg kg⁻¹) | 94.3 | 103 | 94.0 | 113 | 184 | 0.7 | 8.4 | 4.5 | 15.4 |
| CV (%) | 9.9 | 3.6 | 4.7 | 2.3 | 2.0 | 1.7 | 1.2 | 1.3 | 1.4 |
| Ni (mg kg⁻¹) | – | 13.3 | 14.6 | 11.7 | 12.9 | 0.4 | 0.7 | 0.8 | 0.7 |
| CV (%) | – | 1.9 | 1.4 | 1.5 | 0.5 | 4.2 | 7.7 | 1.5 | 14.0 |
| Pb (mg kg⁻¹) | 280 | 229 | 244 | 2.7 | 2.6 | <0.2 | 0.2 | 0.8 | 0.7 |
| CV (%) | 3.1 | 2.2 | 2.5 | 4.3 | 4.4 | – | 15.8 | 4.3 | 23.0 |
| Zn (mg kg⁻¹) | 498 | 422 | 483 | 509 | 609 | 5.3 | 10.6 | 16.1 | 11.8 |
| CV (%) | 1.9 | 4.4 | 5.8 | 0.7 | 1.0 | 5.7 | 0.5 | 2.0 | 1.7 |

and Mn (+63%). For calcium chloride and ammonium acetate extraction the agreement between the two sets of analyses was poor. These differences could probably be largely attributed to analytical imprecision since the extracted contents for these extractants approached the detection limits of flame atomic absorption spectrometry. It was concluded that the temporal stability of extractable contents was acceptable only for EDTA and acetic acid extraction. It should also be remembered in assessing these results that some of the variability found over these long periods will not be due entirely to ageing of the sample but to changes in laboratory staff, in analytical practices, in stability of calibrating solutions, in the temperature of extraction[71] or in the speed of rotation of the mechanical shaker used for the extraction.

### 3.2.2 Effect of Speed of Rotation of Mechanical Shaker

The effects of speed of rotation of the mechanical shaker on soil extractable contents was briefly studied by carrying out extractions with acetic acid and with ammonium acetate using two end-over-end shakers, one operated at 28 rpm and the other at 44 rpm. The acetic acid extractable contents obtained with the more rapid shaker for seven elements increased by an average of 11% (range 4 to 23%) while that of manganese decreased by 22%. For ammonium acetate extraction, the contents of Cd, Cu, Mn and Zn also increased (mean 11%) with the speed of

**Table 4** *Stability tests. Percentage changes on trace element extractable amounts*

| Extract | Cd | Cr | Cu | Fe | Mn | Ni | Pb | Zn |
|---|---|---|---|---|---|---|---|---|
| EDTA (a) | −3.5 | −31 | +5.0 | +0.2 | −8.7 | +5.2 | +6.5 | +14.0 |
| EDTA (b) | +19 | −2.8 | +2.6 | – | −0.3 | – | −12.8 | −3.0 |
| HOAc (a) | +7.4 | +36 | +21 | −18 | +63 | +10.2 | +1.9 | +20 |
| CaCl$_2$ (a) | +118 | – | +158 | – | +1120 | +53 | – | +100 |
| NH$_4$Ac (a) | +7.5 | +101 | +77 | +131 | +242 | −11.3 | −19 | −27 |

(a) percentage change from 1989 to 1990
(b) percentage change from 1986 to 1990

rotation. For all the elements, excepting iron, the increase ranged from 2 to 51% with a mean increase of 20%.

It was further concluded that, in future trials of temporal stability, closer control of these other factors was desirable, that a longer period of study was required and more frequent sampling over the period of study so that it can be determined whether or not a plateau region of temporal stability of extractable contents is reached. The advisability and practicability of storage of soil at 4°C, instead of room temperature, during the trial should be considered particularly with regard to Mn. For some extractants and some elements, it was also obvious that methods of analysis, such as electrothermal atomic absorption spectrometry, that are more sensitive than either flame atomic absorption or inductively coupled plasma emission spectrometry were required. This carries with it the possible need for clean laboratory conditions if contamination is to be avoided.

## 3.3 First Interlaboratory Studies

### 3.3.1 The Participants in the Working Group

Each participant had a wide range of experience in the determination of trace metal partitioning using single extraction schemes. Each laboratory was requested by BCR to accept the firm commitment to remain within the working group for the duration of the programme in order to maintain the continuity of the work and to fulfil the criteria of the learning process and the certification exercise. The intercomparisons were carefully designed to identify the particular sources of error in the methods used by each laboratory and to minimise or eliminate them. The BCR supplied calibrant solutions and supported the preparation and distribution of the samples. The BCR also supported the travel to technical meetings. Each member of the group agreed to submit the results for each exercise within the programme timetable. Support for each laboratory for the subsequent exercises in the programme was based on keeping this commitment.

### 3.3.2   Preparation and Characterisation of the First Test Material

#### 3.3.2.1   Sample Preparation

Sludge-treated soil was air-dried at 25 °C and sieved to provide some 15 kg of dry soil (particle size < 2 mm). The residue particles ( > 2 mm) represented only 2.6% of the whole sample. Loss on ignition of the dry soil at 450 °C was 20.5%. The < 2 mm soil sample was homogenised by putting it into a large polyethylene bag and rolling it in the bag. The whole sample was then poured on to a clean polyethylene sheet, mixed, coned and quartered. The sample was split into two halves by bulking opposite quarters. One of the halves was set aside and the remaining half was further coned and quartered, opposite quarters bulked, one portion set aside and the remaining portion (one-quarter of the whole) was again coned and quartered. Bottles (brown glass) were then filled alternatively from opposite quarters of this portion using a nylon spatula. Empty and filled bottles were weighted at regular intervals to ensure that there was at least 100 g in each bottle. This process was repeated with the other two opposite quarters until all the material had been bottled. The other quarter of the whole sample was dealt with in the same way. The other half of the whole sample which had been set aside was also sampled and bottled to give, finally, a grand total of 125 bottles each containing *ca.* 100 g soil.

#### 3.3.2.2   Homogeneity Testing

Five bottles taken out during the bottling procedure were used for homogeneity testing of the material. From each of these five bottles of soil, representative subsamples of 5 g and 10 g were taken by coning and quartering. Each of these subsamples was extracted with EDTA 0.05 mol $L^{-1}$ solution (ammonium salt, pH 7), at a solution-to-soil ratio of 5, in an end-over-end shaker for 1 h at room temperature (approximately 18 to 23 °C). Similar extractions were carried out overnight (*ca.* 16 h) with ammonium acetate 1 mol $L^{-1}$ solution (pH 7) at a solution-to-soil ratio of 16. These extracts were analysed for Cu, Mn, Ni and Zn, with the results shown in Table 5.

From Table 5, it can be seen that the combined reproducibility of sampling and analysis was very similar both for 10 g and 5 g sample masses. Thus for EDTA, the mean CV for four elements was 2.0% for a 10 g sample and 2.8% for a 5 g sample; for $NH_4OAc$, the means were respectively 6.2 and 4.5%. The reproducibility was, however, slightly poorer with $NH_4OAc$ extraction than with EDTA, as might be expected at the much lower concentrations extracted by the former. The mean values of the EDTA extract concentrations were 10, 11 and 5% lower for copper, manganese and nickel when a 5 g soil mass was used than those obtained when a 10 g mass was used. With zinc the mean obtained with a 5 g intake was 5% greater than with a 10 g intake. It was concluded that the samples were homogeneous enough for sampling at 5 g or 10 g soil masses to be representative and that any bias in the absolute concentration values obtained at these two masses was in general unlikely to exceed 10%.

**Table 5** *Homogeneity testing using 10 g and 5 g sample masses. Summary of mean extractable contents and CVs for EDTA and ammonium acetate extracts of soil*

| | EDTA 0.05 mol $L^{-1}$ | | | | $NH_4OAc$ 1 mol $L^{-1}$ | | | |
|---|---|---|---|---|---|---|---|---|
| | 10 g | | 5 g | | 10 g | | 5 g | |
| Element | Mean mg kg$^{-1}$ | CV % | Mean mg kg$^{-1}$ | CV % | Mean mg kg$^{-1}$ | CV % | Mean mg kg$^{-1}$ | CV % |
| Cu | 134 | 2.4 | 120 | 0.6 | 5.3 | 6.0 | 5.0 | 3.4 |
| Mn | 87 | 2.4 | 77 | 4.8 | 12.7 | 2.9 | 12.6 | 1.9 |
| Ni | 13.5 | 1.4 | 12.8 | 3.1 | 0.88 | 9.5 | 1.04 | 8.6 |
| Zn | 409 | 1.7 | 431 | 2.8 | 12.0 | 6.3 | 12.4 | 4.2 |
| Mean CV | | 2.0 | | 2.8 | | 6.2 | | 4.5 |

Further homogeneity testing was carried out with five replicate 10 g samples taken from five different bottles, extracted with EDTA and $NH_4OAc$ and analysed by AAS for the elements Cd, Cr, Cu, Fe, Mn, Ni, Pb and Zn. At the same time five replicate analyses were carried out on 10 g samples from one bottle. The mean value obtained for the between-bottle analysis of EDTA extracts for each element was in close agreement with the within-bottle mean value. Similarly the ammonium acetate between-bottle and within-bottle extracts have very similar mean concentrations. The overall between-bottle mean CV for the eight elements in the EDTA extracts was 2.2% compared with a within-bottle mean CV of 0.9%. For ammonium acetate extraction the respective mean CVs were 7.8% and 4.1%. The expected deterioration in reproducibility on going from the analysis of samples from a single bottle to that of separate samples from five different bottles extracted individually is, therefore, relatively small, averaging about a factor of two. It was concluded that there was no trend or bias in the results of the analysis of samples taken from the various bottles for EDTA and $NH_4OAc$ extractable amounts for the eight elements. It was further concluded that the soil sample was homogeneous and has been representatively subsampled into the 100 g bottle quantities and that representative subsamples of 5 g or 10 g can be taken from the bottles for analysis.

### 3.3.2.3 Temporal Stability of Soil Extractable Contents

In the work carried out in the preliminary studies, it was satisfactorily established that:

(a) EDTA extractable contents are temporally stable for the elements Cd, Cr, Cu, Fe, Mn, Ni, Pb and Zn over both one and three-year periods;
(b) Acetic acid extractable contents are acceptably stable for Cd, Cu, Fe, Ni, Pb and Zn but more variable than EDTA extraction over time;
(c) Calcium chloride extractable contents are not acceptably stable with time or, alternatively, FAAS is not sufficiently sensitive to be able to determine the

low element contents extracted with precision;

(d) Ammonium acetate extractable contents were stable for the elements Cd and Ni, acceptably stable for Pb and Zn but unacceptable for Cu, Cr, Fe and Mn.

### 3.3.3   Extraction Procedures for the First Interlaboratory Trials

The extraction schemes (in their final version) are given in Appendix 1. All participants in the interlaboratory studies were requested to undertake the analyses according to the prescribed procedures. Five replicate determinations (*i.e.* five independent extractions of soil subsamples) had to be carried out for EDTA 0.05 mol $L^{-1}$. The extracts had to be analysed in triplicate by AAS or ICP for the elements Cd, Cr, Cu, Ni, Pb and Zn. On a single blank extract, *i.e.* with no soil, three replicate determinations of these six elements were to be carried out and reference solution A was to be analysed in triplicate for the same six elements. A further 5 g soil sample was to be taken at the same time and dried in an oven at 105 °C for 2 h to determine the correction to the dry mass basis used for the analyses. The same analyses had to be undertaken (optionally) for ammonium acetate 1 mol $L^{-1}$ and corresponding reference solution B, and acetic acid 0.43 mol $L^{-1}$ and corresponding reference solution C.

The procedure for soil extraction prescribed the use of an end-over-end mechanical shaker but, to determine the extent to which the type of shaker affected the extractable concentrations, laboratories were invited to repeat their extractions and analyses with any other type of shaker available. In this instance analysis of triplicate extractions were considered to be sufficient. The type of shaker, its speed of oscillation or rotation and the temperature of the extract in the extracting vessel at the end of the extraction period were requested to be reported.

### 3.3.4   Results of the First Interlaboratory Trial

Details of the contributions of the different laboratories (methods of determination, type of shaker, temperature, *etc.*) are reported elsewhere,[62] along with the individual results obtained. Results obtained from EDTA extractions were normalised against results of calibrant solutions to take into account possible calibration errors. Systematic errors were detected in some laboratories, exclusively due to calibration errors, which illustrated the care to be detected for ensuring accurate determinations of the whole analysis; any efforts to compare extraction performances are useless if calibration errors are the main cause of discrepancies. The results for the EDTA extracts of soil showed that, for all the elements except Cr, interlaboratory agreement within about $\pm 10\%$ was obtainable using current methodologies and not a very rigorous criterion for the rejection of outliers (Table 6). It was concluded, however, that an improvement in the determination of Cr and avoidance of contamination would be required before a soil could be certified for EDTA-extractable contents of this element.

In the case of acetic acid results, interlaboratory agreement for Cd, Cr, Cu and Zn was more than adequate for certification to be attempted (Table 7). For the other two elements, Ni and Pb, a more rigorous selection of analytical values or a more sensitive analytical technique in the case of Pb might be required for successful certification.

With respect to the ammonium acetate results, indications of contamination for Cr, Cu and Ni, and the significant level of blank concentrations indicated that improvements in the analytical and the other technologies adopted were required before certification of soils for ammonium acetate extractable contents will be possible (Table 8).

The blank levels encountered by most laboratories were of little importance in their effects on the analyses of the EDTA extracts except for one isolated value for Zn. In the case of the analyses of the acetic acid extracts the blanks again presented little problem except for Pb and again for one isolated case for Zn. For the ammonium acetate extracts, important blank values were found for Ni and Pb and in one case for Cr, reflecting the fact that the concentrations of these elements were low. Blank values for Cu were also not negligible. From this aspect too, the difficulties in making use of ammonium acetate extracts had not yet been overcome.

### 3.3.5 Second Interlaboratory Trial on Calcareous Soil

#### 3.3.5.1 Selection of Extractants

Three different extractants were discussed for the feasibility study on calcareous soil, namely EDTA, DTPA and mixed acid ammonium acetate/EDTA. EDTA

**Table 6** *Summary of EDTA soil extract results. Overall means of laboratory raw mean contents and CVs. Overall means of corrected (calibration and blank) laboratory means (mean values in mg kg⁻¹ extractable contents on a dry matter basis)*

|                     | Cd   | Cr  | Cu  | Ni   | Pb  | Zn  |
|---------------------|------|-----|-----|------|-----|-----|
| Mean of raw means   | 23.1 | 8.1 | 162 | 16.3 | 255 | 492 |
| St. deviation       | 2.1  | 2.1 | 14  | 2.1  | 30  | 29  |
| CV %                | 9.2  | 26  | 8.5 | 13   | 12  | 5.8 |
| Mean of corr. means | 23.8 | 7.9 | 170 | 16.5 | 257 | 493 |

**Table 7** *Summary of acetic acid soil extract results. Overall means of laboratory raw mean contents and CVs. Overall means of corrected (calibration and blank) laboratory means (mean values in mg kg⁻¹ extractable contents on a dry matter basis)*

|                     | Cd   | Cr   | Cu   | Ni   | Pb  | Zn  |
|---------------------|------|------|------|------|-----|-----|
| Mean of raw means   | 19.3 | 26.1 | 29.2 | 15.7 | 3.4 | 522 |
| St. deviation       | 1.5  | 2.2  | 2.9  | 2.9  | 0.8 | 36  |
| CV %                | 7.5  | 8.2  | 10   | 18   | 25  | 6.9 |
| Mean of corr. means | 20.2 | 25.5 | 29.5 | 15.9 | 3.4 | 538 |

**Table 8**   *Summary of ammonium acetate soil extract results. Overall means of laboratory raw mean contents and CVs. Overall means of corrected (calibration and blank) laboratory means (mean values in mg kg$^{-1}$ extractable contents on a dry matter basis)*

|                     | Cd   | Cr   | Cu   | Ni   | Pb   | Zn   |
|---------------------|------|------|------|------|------|------|
| Mean of raw means   | 3.43 | 1.39 | 5.65 | 1.42 | 2.21 | 18.4 |
| St. deviation       | 0.37 | 0.56 | 1.32 | 0.32 | 0.59 | 4.2  |
| CV %                | 10.9 | 41   | 23   | 23   | 27   | 23   |
| Mean of corr. means | 3.58 | 1.33 | 5.57 | 1.49 | 2.18 | 18.6 |

was assumed to extract both carbonate-bound and organically-bound fractions of metals and was hence considered to be suitable for calcareous soil analysis. Mixed acid ammonium acetate/EDTA reagent was also discussed but this method was discarded as there is some evidence that EDTA at pH 5.5 can precipitate Cr, Pb and Zn as was observed in polluted sediments.[72] Moreover, the benefits of supplementing the acid ammonium acetate did not seem worth the more complicated procedure. Finally, DTPA extraction was stressed to be more complicated than the EDTA one and was recognised to be often misused;[73] in addition, DTPA extracts less than EDTA which might lead to sensitivity problems. This method was, however, retained for the feasibility study owing to its high degree of acceptance. EDTA was the method of preference as it was extensively tested in previous studies.

### 3.3.5.2   Interlaboratory Study on Calcareous Soil

It was decided to organise an interlaboratory study using EDTA and DTPA as extraction schemes for the determination of extractable trace metal contents in sewage sludge-amended calcareous soil. The same procedure as the one described before was followed for the organisation of the trial, involving the same laboratories in the working group. A sewage-contaminated calcareous soil was selected from the bank of reference materials of the Environment Institute of the Joint Research Centre of Ispra (Italy) in order to present both the characteristics of a calcareous soil (CaCO$_3$ content of 228 g kg$^{-1}$) with high heavy metal contents. The soil sample was composed of 15.4% sand (2 mm to 50 $\mu$m), 9.3% coarse silt (50 to 20 $\mu$m), 34.0% fine silt (20 to 2 $\mu$m) and 41.3% clay (<2 $\mu$m). This material was sieved to 2 mm, homogenised, sterilised and bottled in dark brown bottles. As already stressed, the most critical steps discussed in the technical scrutiny of the results of the interlaboratory study were the type and speed of shaking, filtration and centrifugation. As a general remark, it was stressed that the use of standard additions for calibration was a prerequisite for electrothermal absorption spectrometry. It was also recalled that the use of an end-over-end shaker is required and, therefore, the data obtained using a horizontal shaker were rejected. The DTPA extraction was criticised owing to the high mass/volume ratio, which limited the volume of extract collected after centrifugation. This procedure was not considered to be applicable to Cr determination.

Table 9 gives the coefficients of variation between laboratories obtained before the technical scrutiny (*i.e.* including all sets of data). As shown in this table, the overall variability in the results for EDTA and DTPA-extractable contents Cd and Cu were comparable. However, the variability observed for the DTPA-extractable contents Ni, Pb and Zn appeared significantly higher in comparison with the EDTA-extractable contents. On the basis of the results obtained in this exercise, the choice of EDTA and DTPA for certification was discussed. Whereas EDTA was widely accepted, the choice of DTPA was more criticised because of its operational difficulties; the wide use of the latter would, however, justify the certification of DTPA-extractable trace element contents, providing that its limitations in comparison with EDTA were clearly identified.

### 3.3.6 Conclusions of the First Interlaboratory Trials

The principal conclusions drawn by the working group from the interlaboratory trials were:

1. The analytical procedures were sufficiently adopted and were sufficiently accurate and reproducible; in the case of EDTA 0.05 mol L$^{-1}$ extracts of sludge-amended soil for the elements Cd, Cu, Ni, Pb and Zn for the preparation of reference materials (including calcareous soil) certified for extractable contents were attempted with good prospects of success. For Cr some improvement in the procedures would be necessary or a more rigorous policy adopted for the rejection of outliers.
2. The results for acetic acid 0.43 mol L$^{-1}$ extracts for the elements Cd, Cr, Cu and Zn also showed that reference materials could probably be prepared with certified extractable contents of these metals. Some modest improvement would be required for Ni and Pb determinations.
3. While Cd results for the analysis of ammonium acetate 1 mol L$^{-1}$ (pH 7) would be acceptable for certification purposes, all the other elements would require considerable improvement for such a purpose and this scheme was not selected for certification.
4. The DTPA extraction used for calcareous soil analysis was criticised owing to the high mass/volume ratio, which limited the volume of extract collected after centrifugation. This procedure was not considered to be applicable to Cr determination. The wide use of this procedure, however, justified its selection

**Table 9** *Coefficients of variations of the means of laboratory means of EDTA and DTPA-extractable contents as obtained in the first interlaboratory study*

| CV(%) | Cd | Cr | Cu | Ni | Pb | Zn |
|---|---|---|---|---|---|---|
| EDTA extracts | 29 | 30 | 24 | 28 | 21 | 27 |
| DTPA extracts | 26 | 79 | 24 | 39 | 35 | 48 |

for the certification of DTPA-extractable trace element contents in a calcareous soil.

# 3.4    Certification of Soil Reference Materials

## 3.4.1    Preparation of the Reference Materials

### 3.4.1.1    CRM 483

The material was collected from Great Billings Sewage farm (Northampton) in
1991.[74,75] Some 300 kg of field-moist soil was collected by multiple sampling to a
depth of 10 cm and bulked into polyethylene bags for transport to the Macaulay
Land Use Research Institute (Aberdeen, UK). The whole soil was air-dried at
30°C for 3 weeks on paper-lined aluminium trays. The dried material was then
gently rolled with a wooden roller to break up large aggregates, sieved through a
2 mm round-hole sieve and stored in tightly-sealed polyethylene bags. The soil
sample was thoroughly mixed and homogenised by rolling on a clean polyethylene sheet for 3 days with occasional mixing by hand. The whole sample was then
gently poured on to a clean polyethylene sheet, mixed and coned and quartered
by hand. The initial sample, nominally 150 kg of air-dry (<2 mm) soil was split
by coning and quartering, bulking opposite quarters to form the half samples,
and setting one half sample aside. The remaining half sample was again coned
and quartered. The coning and quartering procedure continued (six cycles) until
the half-sample weight was approximately 2 kg. From opposite quarters of this
half-sample 20 subsamples were taken alternately by nylon spatula into pre-
cleaned brown glass bottles (capped by polyethylene screwcaps). Each bottle
contained approx. 70 g and a total of 1280 bottles were obtained. 128 bottles (two
from each final half-sample) were set aside for homogeneity and stability testing.

### 3.4.1.2    CRM 484

The sampling of the material (terra rossa soil) was carried out in 1991 in a farm
plot amended with sewage sludge from a water treatment centre located in
Northeast Catalonia, Spain.[74,75] Samples were taken from an area of about
250 m² with a small shovel to a depth of about 10 cm and sifted on-site by hand
through a 0.5 cm nylon sieve into polyethylene bags. The samples were taken to
the Water Treatment Centre and again sieved through a 20 cm diameter nylon
sieve with a mesh size of 2 mm into polyethylene bags for transport to the
laboratory of Analytical Chemistry of the University of Barcelona. The soil was
then spread over a polyethylene sheet and air-dried at 30°C for one week to final
water content of 1.5%. The air-dried soil was packed into a 100 litre polyethylene
container, tightly sealed and dispatched to the Environment Institute of the Joint
Research Centre of Ispra (Italy) for homogenisation and bottling. The air-dried
(<2 mm) soil sample was transferred in total (91 kg) into a mixing drum filled
with dry argon and placed on a roll-bed capable of handling 100 kg samples. The
homogenisation of this soil, with its large spread of particle sizes, from just below
2 mm down to fractions of a micrometer, required particular care. This pro-

cedure was, therefore, carried out by mixing in the drum for over 4 weeks. The bottling procedure was performed as follows: to prevent segregation of fine particles, 10 samples were taken from the centre of the drum immediately upon stopping the rotation of the mixing drum, and were placed into 10 pre-cleaned brown glass bottles, so each contained a minimum of 70 g of soil. The drum was again rotated for a further two minutes and a further 10 samples were subsampled in the same way into bottles. The subsampling and bottling operation was continued until 1000 bottles of the soil were obtained. 100 bottles, selected sequentially over the whole bottling procedure were sent to the Macauley Land Use Research Institute for homogeneity and stability testing.

## 3.4.1.3 CRM 600

The material was collected at San Pellegrino Parmense (Italy) in February 1994, following a prospective study of various sites in Italy, which aimed at identifying a material with reasonably high calcium carbonate content.[76,77] About 250 kg of soil was collected by shovelling from the surface to a depth of 10 cm (collection of top layers). Stones and large plant litter were removed prior to sieving with a 2 mm mesh. The fraction less than 2 mm was collected in stainless-steel trays in which the material was dispersed in thin layers of a few cm of thickness to dry at ambient temperature. The material was sieved again after drying to remove lumps which formed during the drying process. The residual moisture content at this stage was found to be 3.8% (measured by taking a separate portion of 1 g dried at 105 °C until constant mass was attained). The sieved material was transferred into a PVC mixing drum filled with dry argon, and was homogenised for 12 days at about 48 rpm. The final material was manually bottled in brown glass bottles. The bottling procedure was carried out by filling 10 bottles, closing the drum and mixing the material again for 2 minutes before bottling another 10 bottles, and so on until only a few cm of soil remained in the drum (which were discarded). All bottles were closed with an insert and a screw cap and stored at ambient temperature. 1050 bottles each containing about 70 g were produced.

## 3.4.1.4 BCR 700

The material is an organic-rich soil which was selected from the bank of Eurosoils of the Joint Research Centre of Ispra.[78,79] The sampling of the material was carried out in 1998 at Hagen, Germany. Samples were taken from an area of about 200 m$^2$ by the Environment Institute of the Joint Research Centre of Ispra (Italy) for homogenisation and bottling. The collected material was picked over for the removal of stones, litter and other material extraneous to soil and exposed to air-drying at ambient temperature. Following air-drying, the soil aggregates were crushed and the soil passed over a 2 mm sieve. The fraction >2 mm was discarded. The soil fraction (91 kg) was transferred into a mixing drum filled with dry argon and placed on a roll-bed capable of handling 100 kg samples. The homogenisation of this soil, showing a large spread of particle sizes, from just below 2 mm down to fractions of a micrometer, required particular care. Mixing was therefore extended to 4 weeks. The bulk homogeneity of the material was tested by taking subsamples from the drum and analysing them by X-ray fluorescence spectrometry. As the data analysis offered no indication for material inhomogeneity, the sample was bottled. The bottling procedure was performed as follows: to prevent segregation of fine particles, 10 samples were taken from the centre of the drum immediately upon stopping the rotation and were placed into 10 pre-cleaned brown glass bottles so each contained a minimum of 70 g of soil. The drum was again rotated for a further two minutes and a further 10 samples were subsampled in the same way. The subsampling and bottling operation was continued until 1200 bottles of the soil were obtained. 100 bottles, selected sequentially over the whole bottling process were selected for homogeneity and stability testing.

## 3.4.2  Homogeneity Studies

The extractants (0.05 mol L$^{-1}$ EDTA, 0.43 mol L$^{-1}$ acetic acid and 0.005 mol L$^{-1}$ DTPA) were prepared as laid out in Appendix 1 at the end of this chapter. All precautions were taken to avoid contamination during the extraction procedures. The trace element contents (Cd, Cr, Cu, Ni, Pb and Zn) in the extracts were determined by inductively coupled plasma atomic emission spectrometry (ICPAES) or electrothermal atomic absorption spectrometry. Calibrant solutions were prepared from stock solutions (1000 mg L$^{-1}$) of the individual elements in 0.5 mol L$^{-1}$ HNO$_3$.

For the homogeneity study, the six elements were determined in each candidate CRM by analysing 10 subsamples taken from one bottle (within-bottle homogeneity test, CV$_W$) and one subsample in each of 20 different bottles selected during the bottling procedure (between-bottle homogeneity test, CV$_B$), strictly following the single extraction procedure requirements. The CVs and the uncertainty U$_{CV}$ for the extractable trace element contents between and within bottles are given in Table 10.

For most of the extractable metal contents in the four soils, the overlap was within the uncertainty U$_T$ of the CV (an approximation of the uncertainty U$_{CV}$ of the CV is calculated as follows: U$_{CV} \approx$ CV$/\sqrt{2n}$). The small difference between the within-bottle and between-bottle CVs is rather an analytical artefact than an indication of inhomogeneity.

In the case of the candidate CRM 483, little analytical difficulty was experienced as illustrated by the good agreement obtained between the within-bottle and between-bottle CVs. For the candidate CRM 484, however, lower extractable contents closer to the detection limits and consequent poorer analytical precision was observed in particular for Cr (EDTA-extractable contents), Cd and Pb (acetic acid-extractable contents). An F-test has been used to test for a significant difference between the within-bottle and between-bottle test results, showing that, with the exception of two results for chromium the variance ratios are well below 1.56, *i.e.* there was no significant difference between the within-bottle and between-bottle test results.

On the basis of these results, the materials were considered to be homogeneous

**Table 10** *Results of the homogeneity tests carried out on the candidate soil samples*

| CRM 483 | Cd | Cr | Cu | Ni | Pb | Zn |
|---|---|---|---|---|---|---|
| EDTA | | | | | | |
| $CV_B$ | 5.7±1.3 | 7.0±1.6 | 5.8±1.3 | 5.7±1.3 | 5.3±1.2 | 5.6±1.3 |
| $CV_W$ | 5.2±1.6 | 8.5±2.7 | 5.4±1.7 | 5.2±1.6 | 5.1±1.6 | 5.1±1.6 |
| | | | | | | |
| Acetic acid | | | | | | |
| $CV_B$ | 1.9±0.4 | 1.4±0.3 | 1.0±0.2 | 2.8±0.6 | 11.0±2.5 | 2.6±0.6 |
| $CV_W$ | 1.6±0.5 | 0.8±0.3 | 1.0±0.3 | 2.8±0.9 | 14.0±4.4 | 2.8±0.9 |
| | | | | | | |
| **CRM 484** | Cd | Cr | Cu | Ni | Pb | Zn |
| EDTA | | | | | | |
| $CV_B$ | 5.0±1.1 | 11.0±2.5 | 5.7±1.3 | 4.3±1.0 | 4.8±1.1 | 5.4±1.2 |
| $CV_W$ | 7.1±2.2 | 6.4±2.0 | 8.0±2.5 | 8.7±2.8 | 8.2±2.6 | 8.2±2.6 |
| | | | | | | |
| Acetic acid | | | | | | |
| $CV_B$ | 5.3±1.2 | 10.2±2.3 | 3.4±0.8 | 7.2±1.6 | – | 6.8±1.5 |
| $CV_W$ | 12.9±4.1 | 10.7±3.4 | 4.5±1.4 | 8.3±2.6 | 4.5±1.4 | 6.8±2.2 |
| | | | | | | |
| **CRM 600** | Cd | Cr | Cu | Ni | Pb | Zn |
| EDTA | | | | | | |
| $CV_B$ | 1.9±0.4 | 10.8±2.4 | 2.4±0.5 | 2.3±0.5 | 7.3±1.6 | 3.1±0.7 |
| $CV_W$ | 2.3±0.5 | 6.9±1.5 | 2.4±0.5 | 2.5±0.6 | 4.5±1.0 | 3.1±0.7 |
| | | | | | | |
| DTPA | | | | | | |
| $CV_B$ | 12.9±2.9 | 5.6±1.2 | 1.7±0.4 | 1.4±0.3 | 1.8±0.4 | 2.3±0.5 |
| $CV_W$ | 1.8±0.4 | 4.0±0.9 | 1.3±0.3 | 1.7±0.4 | 2.0±0.4 | 1.4±0.3 |
| | | | | | | |
| **BCR 700** | Cd | Cr | Cu | Ni | Pb | Zn |
| EDTA | | | | | | |
| $CV_B$ | 2.9±0.5 | 5.3±0.8 | 2.9±0.5 | 3.6±0.6 | 3.2±0.5 | 2.6±0.4 |
| $CV_W$ | 2.6±0.8 | 5.3±1.7 | 2.8±0.9 | 3.3±1.0 | 4.7±1.5 | 2.9±0.9 |
| | | | | | | |
| Acetic acid | | | | | | |
| $CV_B$ | 2.0±0.3 | 2.6±0.4 | 1.3±0.2 | 2.1±0.3 | 4.8±1.4 | 2.1±0.5 |
| $CV_W$ | 2.3±0.7 | 2.1±0.7 | 1.0±0.3 | 1.9±0.6 | 4.4±1.4 | 1.7±0.5 |

at a level of 5 g (as specified in the EDTA protocol for the four soils and acetic acid extraction protocol for the CRMs 483 and 484 and BCR 700) and 10 g (as specified in the DTPA extraction protocol for the CRM 600).

### 3.4.3 Stability Studies

The stability of the extractable trace element contents was tested to determine the suitability of the soils as reference materials. For the three CRMs 483, 484 and 600, sets of bottles were kept at different temperatures, namely −20°C, +20°C and +40°C and EDTA, acetic acid and DTPA-extractable contents of Cd, Cr, Cu, Ni, Pb and Zn were determined (in five replicates) after 1, 3, 6 and 12 months. The procedures used were the same as in the homogeneity study. No instability was detected at +20°C and +40°C, using the values obtained at −20°C as reference; problems likely due to changes in extractability upon defreezing of the materials were, however, noted and it was recommended not to store materials at temperatures below $+4\,°C$.

In the case of CRM 600, however, values obtained for Cu clearly demonstrated that instability was likely both for the EDTA and DTPA-extractable contents; consequently, this element was withdrawn from certification. The variations in the results of DTPA-extractable contents of Pb and Zn clearly indicated that risks of instability were likely for these elements. Consequently, it was decided to withdraw them from certification and, instead, propose indicative values. This study demonstrates that risks of instability may occur at $40\,°C$ which may be due *e.g.* to possible changes in the extractability of some elements (*e.g.* Cu, Pb and Zn); these changes induced by the high storage temperature could be related to changes in the status of the organic matter or in the crystallographic form of Fe or Mn compounds. The stability of Cd contents at $+20\,°C$ was considered to be acceptable for certification whereas the extractable contents of Cu (both for EDTA and DTPA) and the DTPA-extractable contents of Pb and Zn were not found to be sufficiently stable for certification. The variations observed for Ni (EDTA-extractable contents) were attributed to the analytical variability rather than to instability. The other elements were considered to be stable at both $+20$ and $+40\,°C$.

The stability experiments for BCR 700 followed the same approach, excepting that storage was made at $+4\,°C$ (instead of $-20\,°C$), $+20\,°C$ and $+40\,°C$. Also in this case, no instability was detected over a period of 12 months. It is recommended to store the materials at $4\,°C$ in the dark.

### 3.4.4  Analytical Methods Used in the Certification

Each laboratory that took part in the certification exercise was requested to make a minimum of five independent replicate determinations of each element on at least two different bottles of the CRMs on different days, following strictly the extraction protocols described in Appendix 1. The laboratories applied their methods under a suitable quality control regime, verifying their calibrants and checking the validity of the method with the reference materials used in the interlaboratory studies. Details on the techniques used are given in the respective certification reports.[74,76,78]

### 3.4.5  Technical Discussion

At the technical meeting, it was recalled that strict observance of the extraction protocols would be a criterion for considering the results for discussion. An example of possible sources of discrepancies occurring as the result of non-adherence to the protocols was given by one laboratory which used a reciprocating shaker instead of the recommended end-over-end shaker and obtained systematically low results in the certification of CRMs 483 and 484. The repetition of the analyses clearly showed that the error was due to this fact (Table 11).

On the basis of this remark, the results of laboratories using a reciprocating shaker were rejected. The shaker speed was also considered to be an important

**Table 11** *Results (in mg kg$^{-1}$) obtained using a reciprocating and end-over-end shaker, respectively*

| CRM 483 | Cd | Cr | Cu | Ni | Pb | Zn |
|---|---|---|---|---|---|---|
| EDTA | | | | | | |
| Reciprocating | $19.9 \pm 1.0$ | $18.2 \pm 1.9$ | $165 \pm 6$ | $24.3 \pm 1.3$ | $149 \pm 12$ | $478 \pm 15$ |
| End-over-end | $24.6 \pm 1.3$ | $27.5 \pm 1.0$ | $218 \pm 12$ | $30.4 \pm 1.6$ | $232 \pm 10$ | $610 \pm 29$ |
| | | | | | | |
| Acetic acid | | | | | | |
| Reciprocating | $7.4 \pm 0.4$ | $5.5 \pm 0.6$ | $17.6 \pm 1.0$ | $9.9 \pm 0.6$ | $2.37 \pm 0.16$ | $257 \pm 3$ |
| End-over-end | $17.5 \pm 0.3$ | $17.2 \pm 0.2$ | $31.7 \pm 0.6$ | $22.2 \pm 0.8$ | $1.47 \pm 0.16$ | $572 \pm 22$ |
| CRM 484 | Cd | Cr | Cu | Ni | Pb | Zn |
| EDTA | | | | | | |
| Reciprocating | $0.48 \pm 0.03$ | $0.46 \pm 0.03$ | $61.0 \pm 1.3$ | $0.92 \pm 0.06$ | $25.7 \pm 1.3$ | $104 \pm 3$ |
| End-over-end | $0.47 \pm 0.01$ | $0.62 \pm 0.04$ | $85.9 \pm 2.3$ | $1.29 \pm 0.03$ | $41.2 \pm 0.9$ | $146 \pm 3$ |
| | | | | | | |
| Acetic acid | | | | | | |
| Reciprocating | $0.22 \pm 0.03$ | $0.21 \pm 0.02$ | $18.8 \pm 1.8$ | $0.93 \pm 0.14$ | $2.46 \pm 0.11$ | $96 \pm 12$ |
| End-over-end | $0.44 \pm 0.03$ | $0.35 \pm 0.0$ | $30.8 \pm 1.0$ | $1.58 \pm 0.21$ | $1.85 \pm 0.14$ | $172 \pm 11$ |

parameter since it represents one of the factors (along with the shaker type) that condition the maintenance of the samples in suspension during the extraction. The protocol stipulates a speed of 30 rpm and speeds ranging from 20 to 40 rpm were considered to be acceptable. The uses of filtration, centrifugation or centrifugation followed by filtration were all found acceptable.

### 3.4.5.1 EDTA Extracts

In some cases, a good agreement of results of laboratories using a high speed reciprocating shaking and results of the bulk of the participants (using end-over-end shaking) could be observed; as stressed before, these results could, however, not be retained for certification since the extraction protocol was not strictly followed. For Cr in CRM 483, it was considered that FAAS using air/acetylene flame without a releasing agent was not acceptable. Two laboratories using nitrous oxide/acetylene flame were on the high side and in the absence of detailed information on this flame in this matrix, these results were rejected. A high dispersion of results for chromium in CRM 484 did not allow this element to be certified. The mean and standard deviation are given as indicative values.

### 3.4.5.2 Acetic Acid Extracts

No particular problems were experienced with acetic acid for the CRM 483, except for lead for which some results were too close to detection limits, the data sets of which were consequently removed. In CRM 484, as observed for EDTA extracts, the wide dispersion of chromium results could not allow certification; the results were given as indicative values.

## 3.4.5.3   DTPA Extracts

With respect to calcareous soil analysis (CRM 600), there was a general consensus to prefer EDTA because this procedure is easier to apply (better soil-to-solution ratio for EDTA). In addition, it was stressed that EDTA and DTPA extractions are closely correlated which renders questionable the use of both extraction procedures at the same time. It was assumed that EDTA extraction enables a complete extraction to be achieved and mimics the mobility of trace metals from soils; DTPA is widely used in the USA and is rather applied to predict plant uptake. The choice of the extractant is, therefore, closely related to the objective of the study. It should be stressed that the suspicion of an instability for the DTPA-extractable contents of Cu, Pb and Zn did not allow these elements to be certified; the risk of result variability due to a possible instability of the extractable contents for these elements should hence be faced when using DTPA in soil studies. As mentioned in the conclusions of the interlaboratory study, the DTPA extraction procedure was not recommended for the determination of Cr, which was not certified and is given as indicative value. Since doubts were expressed on the stability of EDTA-extractable contents of Cu, and of DTPA-extractable contents of Cu, Pb and Zn, these elements were also not certified and indicative values were proposed.

## 3.4.6   Certified Values

The certified values (unweighted mean of $p$ accepted sets of results) and their uncertainties (half width of the 95% confidence intervals) are given in Tables 12 to 14 as mass fractions of the respective extracts, EDTA, acetic acid and DTPA (based on dry mass) in mg kg$^{-1}$. Indicative (not certified) values are indicated in brackets.

## 3.4.7   Indicative Values

During the course of the certification some laboratories carried out other types of extraction procedures which are described in Appendix 1, *i.e.* calcium chloride, ammonium nitrate and sodium nitrate. These results are given in Tables 15 to 17 together with the standard deviation obtained, the method(s) used and the number of sets of results. It is emphasised that the values of this table were not certified, but were considered of interest to soil scientists.

The improved three-step BCR sequential extraction procedure[34,35] was also applied to CRM 483 to check the suitability of such a procedure for soil analysis in the frame of a small-scale interlaboratory study in which six laboratories participated. Each laboratory that took part in the collaborative exercise to test the sequential extraction procedure on CRM 483 was requested to make a minimum of five independent replicate determinations of each element on at least two different bottles of the CRM on different days, following strictly the modified BCR sequential extraction scheme described in Appendix 1. The participants were also encouraged to determine the *aqua regia*-extractable trace

**Table 12**   *Certified contents of EDTA-extractable trace elements.* p = *no. of accepted sets of results*

|  | Element | Certified value (mg kg$^{-1}$) | Uncertainty (mg kg$^{-1}$) | CV (%) | p |
|---|---|---|---|---|---|
| CRM 483 | Cd | 24.3 | 1.3 | 5.3 | 18 |
|  | Cr | 28.6 | 2.6 | 9.1 | 9 |
|  | Cu | 215 | 11 | 5.1 | 17 |
|  | Ni | 28.7 | 1.7 | 5.9 | 17 |
|  | Pb | 229 | 8 | 3.5 | 17 |
|  | Zn | 612 | 19 | 3.1 | 17 |
| CRM 484 | Cd | 0.51 | 0.03 | 5.9 | 13 |
|  | Cu | 88 | 4 | 4.5 | 14 |
|  | Ni | 1.39 | 0.11 | 7.9 | 15 |
|  | Pb | 47.9 | 2.6 | 5.4 | 16 |
|  | Zn | 152 | 7 | 4.6 | 17 |
| CRM 600 | Cd | 2.68 | 0.09 | 3.4 | 18 |
|  | Cr | 0.206 | 0.021 | 10.2 | 12 |
|  | (Cu) | (57.3) | (2.5) | 4.4 | 17 |
|  | Ni | 4.52 | 0.25 | 5.5 | 11 |
|  | Pb | 59.7 | 1.8 | 3.0 | 17 |
|  | Zn | 383 | 12 | 3.1 | 16 |
| CRM 700 | Cd | 65.2 | 3.2 | 4.9 | 18 |
|  | Cr | 10.2 | 0.80 | 7.8 | 17 |
|  | Cu | 88.7 | 2.7 | 3.0 | 18 |
|  | Ni | 53.1 | 2.6 | 4.9 | 18 |
|  | Pb | 103 | 4.3 | 4.2 | 18 |
|  | Zn | 507 | 18 | 3.6 | 18 |

**Table 13**   *Certified contents of acetic acid-extractable trace elements.* p = *no. of accepted sets of results*

|  | Element | Certified value (mg kg$^{-1}$) | Uncertainty (mg kg$^{-1}$) | CV (%) | p |
|---|---|---|---|---|---|
| CRM 483 | Cd | 18.3 | 0.6 | 3.3 | 18 |
|  | Cr | 18.7 | 1.0 | 5.3 | 17 |
|  | Cu | 33.5 | 1.6 | 4.8 | 18 |
|  | Ni | 25.8 | 1.0 | 3.9 | 15 |
|  | Pb | 2.10 | 0.25 | 11.9 | 12 |
|  | Zn | 620 | 24 | 3.9 | 18 |
| CRM 484 | Cd | 0.48 | 0.04 | 8.3 | 18 |
|  | Cu | 33.9 | 1.4 | 4.1 | 16 |
|  | Ni | 1.69 | 0.15 | 8.9 | 11 |
|  | Pb | 1.17 | 0.16 | 13.7 | 17 |
|  | Zn | 193 | 7 | 3.6 |  |
| CRM 700 | Cd | 67.2 | 2.5 | 3.7 | 19 |
|  | Cr | 18.8 | 1.0 | 5.3 | 19 |
|  | Cu | 35.9 | 1.4 | 3.9 | 19 |
|  | Ni | 98.2 | 4.4 | 4.5 | 19 |
|  | Pb | 4.84 | 0.35 | 7.2 | 15 |
|  | Zn | 713 | 24 | 3.4 | 18 |

**Table 14** *Certified contents of DTPA-extractable trace elements. Indicative (not certified) values are given in brackets. p = no. of accepted sets of results*

|  | Element | Certified value (mg kg$^{-1}$) | Uncertainty (mg kg$^{-1}$) | CV (%) | p |
|---|---|---|---|---|---|
| CRM 600 | Cd | 1.34 | 0.04 | 3.0 | 17 |
|  | (Cr) | (0.014) | (0.003) | 21.4 | 12 |
|  | (Cu) | (32.3) | (1.0) | 3.1 | 16 |
|  | Ni | 3.31 | 0.13 | 3.9 | 14 |
|  | (Pb) | (15.0) | (0.5) | 3.3 | 16 |
|  | (Zn) | (142) | (6) | 4.2 | 17 |

**Table 15** *Indicative values of calcium chloride extractable trace element contents (mg kg$^{-1}$). p = no. of accepted sets of results*

| Element/CRM | Mean | SD | CV (%) | p | Techniques used |
|---|---|---|---|---|---|
| **CRM 483** |  |  |  |  |  |
| Cd | 0.45 | 0.05 | 11.1 | 10 | FAAS, ETAAS, ICPAES |
| Cr | 0.35 | 0.09 | 25.7 | 9 | FAAS, ETAAS, ICPAES |
| Cu | 1.2 | 0.4 | 33.3 | 11 | FAAS, ETAAS, ICPAES |
| Ni | 1.4 | 0.2 | 14.3 | 10 | FAAS, ETAAS, ICPAES |
| Pb | <0.06 | – | – | 8 | FAAS, ETAAS, ICPAES |
| Zn | 8.3 | 0.7 | 8.4 | 9 | FAAS, ETAAS, ICPAES |
| **CRM 484** |  |  |  |  |  |
| Cd | <0.08 | – | – | 9 | FAAS, ETAAS, ICPAES |
| Cr | <0.09 | – | – | 7 | FAAS, ETAAS, ICPAES |
| Cu | 0.67 | 0.29 | 43.3 | 10 | FAAS, ETAAS, ICPAES |
| Ni | <0.05 | – | – | 9 | FAAS, ETAAS, ICPAES |
| Pb | <0.06 | – | – | 7 | FAAS, ETAAS, ICPAES |
| Zn | 0.31 | 0.17 | 54.8 | 7 | FAAS, ETAAS, ICPAES |

element contents (using the ISO standard 11466) in the original sample and in the residue after extraction in order to provide an additional control tool for the quality of analyses (see Appendix 1). Analyses were carried out under a suitable quality control regime, using CRM 601 as a means of verifying the accuracy of the measurements. The methods of final determination were flame atomic absorption spectrometry (FAAS); electrothermal atomic absorption spectrometry (ETAAS) or inductively coupled plasma atomic emission spectrometry (ICPAES). All precautions were taken with respect to the protocol application and the calibration. The entire participants reported their experimental conditions in order to demonstrate the traceability of the results, in particular with respect to the calibrants and calibration procedures used (verified stoichiometry and purity) and the operational conditions related to the sequential extraction scheme. Details on the experimental conditions are available elsewhere.[80,81] All the results submitted by the participating laboratories were discussed at a technical evaluation meeting to confirm the accuracy of the methods of analysis. For each set of results (five replicates), the mean value and the standard deviation

**Table 16** *Indicative values of sodium nitrate-extractable trace element contents* (*mg kg$^{-1}$*). p = *no. of accepted set of results*

| Element/CRM | Mean | SD | CV (%) | p | Techniques used |
|---|---|---|---|---|---|
| **CRM 483** | | | | | |
| Cd | 0.08 | 0.03 | 37.5 | 6 | FAAS, ETAAS, ICPAES |
| Cr | 0.30 | 0.07 | 23.3 | 4 | FAAS, ETAAS, ICPAES |
| Cu | 0.89 | 0.22 | 24.7 | 6 | FAAS, ETAAS, ICPAES |
| Ni | 0.65 | 0.07 | 10.8 | 5 | FAAS, ETAAS, ICPAES |
| Pb | <0.03 | – | – | 4 | FAAS, ETAAS, ICPAES |
| Zn | 2.7 | 0.8 | 29.6 | 5 | FAAS, ETAAS, ICPAES |
| **CRM 484** | | | | | |
| Cd | <0.05 | – | – | 7 | FAAS, ETAAS, ICPAES |
| Cr | <0.03 | – | – | 6 | FAAS, ETAAS, ICPAES |
| Cu | 0.48 | 0.15 | 31.2 | 8 | FAAS, ETAAS, ICPAES |
| Ni | 0.023 | 0.005 | 21.7 | 6 | FAAS, ETAAS, ICPAES |
| Pb | <0.06 | – | – | 7 | FAAS, ETAAS, ICPAES |
| Zn | 0.09 | 0.04 | 44.4 | 6 | FAAS, ETAAS, ICPAES |

**Table 17** *Indicative values of ammonium nitrate-extractable trace elements contents* (*mg kg$^{-1}$*). p = *no. of accepted sets of results*

| Element/CRM | Mean | SD | CV (%) | p | Techniques used |
|---|---|---|---|---|---|
| **CRM 483** | | | | | |
| Cd | 0.26 | 0.05 | 19.2 | 9 | FAAS, ETAAS, ICPAES |
| Cr | 0.27 | 0.10 | 37.0 | 8 | FAAS, ETAAS, ICPAES |
| Cu | 1.2 | 0.3 | 25.0 | 9 | FAAS, ETAAS, ICPAES |
| Ni | 0.65 | 0.30 | 46.2 | 9 | FAAS, ETAAS, ICPAES |
| Pb | 0.020 | 0.013 | 65.0 | 4 | FAAS, ETAAS, ICPAES |
| Zn | 6.5 | 0.9 | 13.8 | 8 | FAAS, ETAAS, ICPAES |
| **CRM 484** | | | | | |
| Cd | 0.003 | 0.002 | 66.7 | 7 | FAAS, ETAAS, ICPAES |
| Cr | <0.06 | – | – | 7 | FAAS, ETAAS, ICPAES |
| Cu | 1.1 | 0.4 | 36.4 | 10 | FAAS, ETAAS, ICPAES |
| Ni | 0.033 | 0.017 | 51.5 | 6 | FAAS, ETAAS, ICPAES |
| Pb | <0.06 | – | – | 7 | FAAS, ETAAS, ICPAES |
| Zn | 0.17 | 0.05 | 29.4 | 9 | FAAS, ETAAS, ICPAES |

were calculated. All the sets of results were found acceptable on technical grounds and so accepted and used for the calculation of the indicative values given in Table 18.

One of the laboratories did not perform *aqua regia* extraction test for pseudo-total determination and for Cd in the residue after sequential extraction. The indicative values were calculated as the arithmetic means of laboratory means (taking into account the number of sets accepted after technical scrutiny) and their standard deviations and the corresponding CV(%) as mass fractions of the respective extracts (based on dry mass) in mg kg$^{-1}$.

**Table 18**   *Indicative values of extractable trace elements contents ($mg\ kg^{-1}$)*
*following the BCR three-step sequential extraction scheme and* aqua
regia *soluble contents.* p = *no. of accepted sets of results*

|            | Element | Mean  | SD   | CV(%) | p | Techniques used |
|------------|---------|-------|------|-------|---|-----------------|
| First step | Cd      | 10.0  | 0.77 | 2.8   | 6 | FAAS, ICPAES |
|            | Cr      | 9.4   | 3.5  | 37    | 6 | ETAAS, ICPAES |
|            | Cu      | 16.8  | 1.5  | 8.9   | 6 | FAAS, ETAAS, ICPAES |
|            | Ni      | 17.9  | 2.0  | 11    | 6 | FAAS, ICPAES |
|            | Pb      | 0.756 | 0.70 | 93    | 6 | ETAAS, ICPAES |
|            | Zn      | 441   | 39   | 8.9   | 6 | FAAS, ICPAES |
| Second step| Cd      | 24.8  | 2.3  | 9.2   | 6 | FAAS, ICPAES |
|            | Cr      | 654   | 108  | 17    | 6 | FAAS, ICPAES |
|            | Cu      | 141   | 20   | 15    | 6 | FAAS, ICPAES |
|            | Ni      | 24.4  | 3.3  | 13    | 6 | FAAS, ETAAS, ICPAES |
|            | Pb      | 379   | 21   | 5.5   | 6 | FAAS, ICPAES |
|            | Zn      | 438   | 56   | 13    | 6 | FAAS, ICPAES |
| Third step | Cd      | 1.22  | 0.48 | 39    | 6 | ETAAS, ICPAES |
|            | Cr      | 2215  | 494  | 22    | 6 | FAAS, ICPAES |
|            | Cu      | 132   | 29   | 22    | 6 | FAAS, ETAAS, ICPAES |
|            | Ni      | 5.9   | 1.4  | 24    | 6 | ETAAS, ICPAES |
|            | Pb      | 66.5  | 22   | 34    | 6 | FAAS, ICPAES |
|            | Zn      | 37.1  | 9.9  | 27    | 6 | FAAS, ICPAES |
| *Aqua regia* | Cd    | 35.9  | 2.9  | 8.1   | 5 | FAAS, ICPAES |
|            | Cr      | 3338  | 541  | 16    | 5 | FAAS, ICPAES |
|            | Cu      | 360   | 11   | 3.1   | 5 | FAAS, ICPAES |
|            | Ni      | 61.2  | 5.6  | 9.2   | 5 | FAAS, ICPAES |
|            | Pb      | 503   | 54   | 11    | 5 | FAAS, ICPAES |
|            | Zn      | 980   | 38   | 3.9   | 5 | FAAS, ICPAES |

Table 19 summarises the data (in mg $kg^{-1}$) for extracted metal and the *aqua
regia* results from the original sample are compared with the sum of the extracted
metals from the three steps plus residual ($\Sigma 3$ steps + *aqua regia* extractable from
residue). No significant differences were observed between the total metal extracted following the *aqua regia* protocol and the sum of extracted metals following
both sequential extraction procedures, which indicates the good quality of the
results obtained.

## 3.5   Conclusions

The stepwise approach followed to select the most suited single extraction
scheme for the determination of trace metal contents in soil and to validate them
in the frame of successive interlaboratory studies has proven to be very efficient
in terms of agreement between the participating laboratories. As described in this
chapter, the interlaboratory studies and the subsequent certifications of soil
reference materials enabled to test in details the EDTA and acetic acid extraction
schemes, as well as DTPA for calcareous soil analysis. As a further development
of the project, the BCR sequential extraction scheme has been applied to CRM
483, showing that this type of procedure can be successfully applied to soil

**Table 19** *Comparison of results obtained following BCR sequential and aqua regia extraction protocols*

| Element (mg kg$^{-1}$) | Cd Mean ± SD | Cr Mean ± SD | Cu Mean ± SD | Ni Mean ± SD | Pb Mean ± SD | Zn Mean ± SD |
|---|---|---|---|---|---|---|
| Fraction 1 | 10.0 ± 0.77 | 9.40 ± 3.5 | 16.8 ± 1.5 | 17.9 ± 2.0 | 0.756 ± 0.70 | 441 ± 39 |
| Fraction 2 | 24.8 ± 2.3 | 654 ± 108 | 141 ± 20 | 24.4 ± 3.3 | 379 ± 21 | 438 ± 56 |
| Fraction 3 | 1.22 ± 0.48 | 2215 ± 494 | 132 ± 29 | 5.9 ± 1.4 | 66.5 ± 22 | 37.1 ± 9.9 |
| Residue | 0.423 ± 0.16 | 183 ± 40 | 43.3 ± 3.8 | 15.2 ± 4.3 | 76.9 ± 17 | 82.1 ± 9.6 |
| Σ3 steps + residual | 36.44 ± 2.5 | 3061 ± 507 | 333 ± 35 | 63.4 ± 5.9 | 523 ± 35 | 998 ± 70 |
| Aqua regia | 36.40 ± 2.8 | 3392 ± 484 | 362 ± 12 | 63.8 ± 7.7 | 501 ± 47 | 987 ± 37 |
| Relative error (%) | 0.12 | -9.75 | -7.98 | -0.63 | 4.42 | 1.13 |

analysis. The availability of four CRMs for EDTA will offer a great support to laboratories which will use this scheme, in terms of method validation and quality control. Three of the four CRMs are also certified for acetic acid and one for DTPA, these two schemes being less popular than EDTA but still widely used.


# 3.6   References

1. G.E. Batley (ed.), *Trace Element Speciation: Analytical Methods and Problems*, CRC Press, Boca Raton, Florida, 1989, Chapter 8.
2. H.A. van der Sloot, L. Heasman and Ph. Quevauviller (eds.), *Harmonization of Leaching/Extraction Tests*, Studies in Environmental Science 70 series, Elsevier, 1997, Chapter 3.
3. G. Rauret, *Talanta*, 1998, **46**, 449.
4. I. Novozamski, Th.M. Lexmon and V.J.G. Houba, *Inter. J. Environ. Anal. Chem.*, 1993, **51**, 47.
5. E. Colinet, H. Gonska, B. Griepink and H. Muntau, EUR Report 8833 EN 1983, 57.
6. A.M. Ure, Ph. Quevauviller, H. Muntau and B. Griepink. *Inter. J. Environ. Anal. Chem.*, 1993, **51**, 135.
7. C.L. Mulchi, C.A. Adamu, P.F. Bell and R.L. Chaney, *Common. Soil Sci. Plant Anal.*, 1992, **23**, 1053.
8. W.L. Lindsay and W.A. Norvell, *Soil Sci. Soc. Am. J.*, 1978, **42**, 421.
9. A. Melich, *Common. Soil Sci. Plant Anal.*, 1984, **15**, 1409.
10. A.M. Ure, R. Thomas and D. Litlejohn, *Inter. J. Environ. Anal. Chem.*, 1993, **51**, 65.
11. S.K. Gupta and C. Aten, *Inter. J. Environ. Anal. Chem.*, 1993, **51**, 25.
12. J.C. Hughes and A.D. Noble, *Common. Soil Sci. Plant Anal.*, 1991, **22**, 1753.
13. C. Juste and P. Solda. *Agronomie*, 1988, **8**, 897.
14. W.P. Pickering, *Ore Geological Reviews*, 1986, **1**, 83.
15. A. Lebourg, T. Sterckeman, H. Cielsielki and N. Proix, *Agronomie*, 1996, **16**, 201.
16. DIN (Deutches Institut für Normung), 1993, Bodenbeschaffenheit, Vornorm, DIN V 19730, in Boden – Chemische Bodenuntersuchungsverfahren, DIN, Berlin, 4.
17. AFNOR (Association Francaisede Normalization), 1994, AFNOR, Paris, 250.
18. UNICHIM (Ente Nazionale Italiano di Unificazione), 1991, UNICHIM, Milan.
19. V.J.G. Houba, I. Novozamski, T.X. Lexmon and J.J. Van der Lee, *Commun. Soil Sci. Plant Anal.*, 1990, **21**, 2281.
20. VSBo (Veordnung über Schadstoffgehalt im Boden), 1986, Nr. 814.12, Publ. eidg. Drucksachen und Materialzentrale, Bern, 1–4.
21. MAFF (Ministry of Agriculture, Fisheries and Food), 1981, Reference Book 427, MAFF, London.
22. J.D.H. Williams, J.K. Syers, R.F. Harris and D.E. Armstrong, *Soil. Sci. Soc. Amer. Proc.*, 1971, **35**, 250.
23. J.D.H. Williams, J.K. Syers, D.E. Armstrong and R.F. Harris, *Soil. Sci. Soc. Amer. Proc.*, 1971, **35**, 556.
24. L.E. Sommers, R.F. Harris, J.D.H. Williams, D.E. Armstrong and J.K. Syers, *Soil. Sci. Soc. Amer. Proc.*, 1972, **36**, 51.
25. A. Tessier, P.G.C. Campbell and M. Bisson, *Anal. Chem.*, 1979, **51**, 844.
26. J. Arunachalam, H. Emons, B. Krasnodebska and C. Mohl, *Sci. Total Environ.*, 1996, **181**, 147.

27. L. Flores, G. Blas, G. Hernández and R. Alcalá, *Water, Air and Soil Pollut.*, 1997, **98**, 105.
28. I. Maiz, M.V. Esnaola and E. Millan, *Sci. Total Environ.*, 1997, **206**, 107.
29. J. Thöming and W. Calmano, *Acta Hydrochim. Hydrobiol.*, 1998, **26**, 338.
30. R.P. Narwal, B.R. Singh and B. Salbu, *Commun. Soil Sci. Plant Anal.*, 1999, **30**, 1209.
31. X. Li, B.J. Coles, M.H. Ramsey and I. Thornton, *Chem. Geol.*, 1995, **124**, 109.
32. L. Orsini and A. Bermond, *Inter. J. Environ. Anal. Chem.*, 1993, **51**, 97.
33. A. Ure, Ph. Quevauviller, H. Muntau, B. Griepink, *Int. J. Environ. Anal. Chem.*, 1993, **51**, 135.
34. A. Sahuquillo, J.F. López-Sánchez, R. Rubio, G. Rauret, R.P. Thomas, C.M. Davidson and A.M. Ure, *Anal. Chim. Acta*, 1999, **382**, 317.
35. G. Rauret, J.F. López-Sánchez, A. Sahuquillo, R. Rubio, C.M. Davidson, A.M. Ure and Ph. Quevauviller, *J. Environ Monit.*, 1999, **1**, 57.
36. A. Barona, I. Aranguiz and A. Elias, *Chemosphere*, 1999, **39**, 1911.
37. J. Szakova, P. Tlustos, J. Balik, D. Pavlikova and V. Vanek, 1999, *Fresenius J. Anal. Chem.*, **363**, 594.
38. E. Diaz-Barrientos, L. Madrid and I. Cardo, 1999, *Sci. Total Environ.*, **242**, 149.
39. R.A. Sutherland and F.M.G. Tack, 2000, *Sci. Total Environ.*, **256**, 103.
40. I.M.C. Lo and X.Y. Yang, *Waste Management*, 1998, **18**, 1.
41. N.D. Kim and J.E. Fergusson, *Sci. Total Environ.*, 1991, **105**, 191.
42. K. Bunzl, M. Trautmannsheimer and P. Schramel, *J. Environ. Qual.*, 1999, **28**, 1168.
43. N.L. Benitez and J.P. Dubois, *Inter. J. Environ. Anal. Chem.*, 1999, **74**, 289.
44. A. Chlopecka, *Water, Air and Soil Pollut.*, 1993, **69**, 127.
45. J. Qian, Z. Wang, X. Shan, Q. Tu, B. Wen and B. Chen, *Environ. Pollut.*, 1996, **91**, 309.
46. A. Chlopecka, *Water, Air and Soil Pollut.*, 1996, **87**, 297.
47. P. Planquart, G. Bonin, A. Prone and C. Massiani, *Sci. Total Environ.*, 1999, **241**, 161.
48. I. Aumada, J. Mendoza, E. Navarrete and L. Ascar, *Commun. Soil Sci. Plant Anal.*, 1999, **30**, 1507.
49. A. García Sánchez, A. Moyano and C. Nuñez, *Commun. Soil Sci. Plant Anal.*, 1999, **30**, 1385.
50. I. Maiz, I. Arambarri, R. Garcia and E. Millan, *Environ. Pollut.*, 2000, **110**, 3.
51. A.S. Jeng and B.R. Singh, *Soil Sci.*, 1993, **156**, 240.
52. L. Ramos, L.M. Hernández and M.J. González, *J. Environ. Qual.*, 1994, **23**, 50.
53. C. Keller and J.C. Vedy, *J. Environ. Qual.*, 1994, **23**, 987.
54. A. Chlopecka, *Sci. Total Environ.*, 1996, **188**, 253.
55. A. Chlopecka, J.R. Bacon, M.J. Wilson and J. Kay, *J. Environ. Qual.*, 1996, **25**, 69.
56. L.Q. Ma and G.D. Rao, *J. Environ. Qual.*, 1997, **26**, 259.
57. R.H.C. Emerson, J.V. Birkett, M. Scrimshaw and J.N. Lester, *Sci. Total Environ.*, 2000, **254**, 75.
58. A.R. Cabral and G. Lefebre, *Water, Air and Soil Pollut.*, 1998, **102**, 329.
59. E.B. Schalscha, P. Escudero, P. Salgado and I. Ahumada, *Commun. Soil Sci. Plant Anal.*, 1999, **30**, 497.
60. A. Barona and F. Romero, *Environ. Technol.*, 1996, **17**, 63.
61. C.N. Mulligan, R.N. Yong and B.F. Gibbs, *Environ. Progress*, 1999, **18**, 50.
62. A.M. Ure, Ph. Quevauviller, H. Muntau, B. Griepink, *EUR Report 14763*, 1993, Office for Official Publications of the European Communities, Luxembourg.
63. A.M. Ure and M.L. Berrow, *Anal. Chim. Acta*, 1970, **52**, 24.
64. M.L. Berrow and G.A. Reaves, *J. Soil Sci.*, 1985, **36**, 31.
65. E. Boken, *Plant and Soil*, 1952, **4**, 154.
66. C.K. Fujimoto and G.D. Sherman, *Soil Sci. Soc. Amer. Proc.*, 1945, **1**, 107.

67. G.F. Berndt, *J. Sci. Food Agric.*, 1988, **45**, 119.
68. R. Bartlett and B. James, *Soil Sci. Soc. Amer. J.*, 1980, **44**, 721.
69. R.J. Haynes and R.S. Swift, *Geoderma*, 1991, **49**, 319.
70. B.T. Warden, *Commun. Soil Sci. Plant Anal.*, 1991, **22**, 169.
71. S.J. Adams and B.J. Alloway, *Environ. Technol. Lett.*, 1988, **9**, 695.
72. M.L. Raisanen, *Analyst*, 1992, **117**, 623.
73. G.A. O'Connor, *J. Environ. Qual.*, 1988, **17**, 715.
74. Ph. Quevauviller, G. Rauret, A. Ure, J. Bacon and H. Muntau, EUR Report 17127, Office for Official Publications of the European Communities, Luxembourg, 1997, (ISBN 92-827-6938-0).
75. Ph. Quevauviller, G. Rauret, R. Rubio, J.F. López-Sánchez, A.M. Ure, J. Bacon and H. Muntau, *Fresenius J. Anal. Chem.*, 1997, **357**, 611.
76. Ph. Quevauviller, M. Lachica, E. Barahona G. Rauret, A. Ure, A. Gomez and H. Muntau, EUR Report 17555, Office for Official Publications of the European Communities, Luxembourg, 1997, (ISBN 92-828-0126-8).
77. Ph. Quevauviller, M. Lachica, E. Barahona, G. Rauret, A. Ure, A. Gomez and H. Muntau, *Sci. Total Environ.*, 1996, **178**, 127.
78. G. Rauret, J.F. López-Sánchez, J. Bacon, A. Gomez, H. Muntau and Ph. Quevauviller, EUR Report 19774, Office for Official Publications of the European Communities, Luxembourg, 2001, (ISBN 92-894-0566-X).
79. M. Pueyo, G. Rauret, J.R. Bacon, A. Gomez, H. Muntau, Ph. Quevauviller and J.F. López-Sánchez, *J. Environ Monit.*, 2001, **3**, 238.
80. G. Rauret, J.F. López-Sánchez, A. Sahuquillo, E. Barahona, M. Lachica, A. Ure, H. Muntau and Ph. Quevauviller, EUR Report 19503, Office for Official Publications of the European Communities, Luxembourg, 2000, (ISBN 92-828-3010-1).
81. G. Rauret, J.F. López-Sánchez, A. Sahuquillo, E. Barahona, M. Lachica, A.M. Ure, C.M. Davidson, A. Gomez, D. Luck, J. Bacon, M. Yli-Halla, H. Muntau and Ph. Quevauviller, *J. Environ Monit.*, 2000, **2**, 228.

# Appendix 1: Extraction Protocols

## EDTA, DTPA and Acetic Acid Extraction Protocols

The procedures to be used in the certification of extractable trace metals (Cd, Cr, Cu, Ni, Pb and Zn) using 0.05 mol L$^{-1}$ EDTA, 0.43 mol L$^{-1}$ acetic acid and 0.005 mol L$^{-1}$ DTPA are described below.

The extraction shall be performed in 250 mL pre-cleaned borosilicate glass, polypropylene or polytetrafluoroethylene (PTFE) bottles using an end-over-end shaker. All laboratory glassware shall be cleaned with 4 mol L$^{-1}$ HNO$_3$ (for at least 30 min), HCl, rinsed with distilled water, cleaned with 0.05 mol L$^{-1}$ EDTA and rinsed again with distilled water.

Extractants shall be prepared according to the following procedure:

1. 0.05 mol L$^{-1}$ EDTA shall be prepared as an ammonium salt solution by adding in a fume cupboard 146.12 $\pm$ 0.05 g of EDTA free acid to 800 $\pm$ 20 mL distilled water and by partially dissolving by stirring in 130 $\pm$ 5 mL of saturated ammonia solution (prepared by bubbling ammonia gas into distilled water). The addition of ammonia shall be continued until all the EDTA has dissolved. The obtained solution shall be filtered through a filter paper of porosity 1.4 to 2.0 $\mu$m (capable of retaining particles of 8.0 $\mu$m size) into a 10 L polyethylene container and diluted with water to 9.0 $\pm$ 0.5 L. The pH shall be adjusted to 7.00 $\pm$ 0.05 by addition of a few drops of either ammonia or hydrochloric acid as appropriate. The solution shall be diluted with distilled water to 10.0 $\pm$ 0.1 L, well mixed and stored in stoppered polyethylene container.
2. 0.43 mol L$^{-1}$ acetic acid shall be prepared by adding in a fume cupboard 250 $\pm$ 2 mL of redistilled glacial acetic acid to about 5 L of distilled water in a 10 L polyethylene container. The solution shall be diluted with distilled water to 10 L volume, well mixed and stored in a stoppered polyethylene container.
3. The DTPA extracting solution shall be prepared containing 0.005 mol L$^{-1}$ diethylenetriamine–pentaacetic acid (DTPA) [C$_{14}$H$_{23}$N$_3$O$_{10}$], 0.01 mol L$^{-1}$ triethanolamine (TEA) [(HOCH$_2$CH$_2$)$_3$N] and adjusted to pH 7.3. To pre-

pare 10 L of this solution, dissolve 149.2 g reagent grade TEA, 19.67 g DTPA
and 14.7 g calcium chloride [$CaCl_2.2H_2O$] in approximately 200 mL distilled
water. Allow sufficient time for the DTPA to dissolve and dilute to approximately 9 L. Adjust the pH to $7.3 \pm 0.5$ with HCl while stirring and dilute to 10
L. This solution is stable for several months.

Extraction shall be batch-wise (*e.g.* shaking), followed by centrifugation or
filtration according to the following procedure:

1. A subsample of 5 g soil shall be transferred to an extraction bottle in which 50
   mL of 0.05 mol $L^{-1}$ EDTA will be added. The obtained mixture shall be
   shaken on an end-over-end shaker operating at 30 rpm for 1 h in a room at
   $20 \pm 2$ °C.
2. A subsample of 5 g soil shall be transferred to an extraction bottle in which
   200 mL of 0.43 mol $L^{-1}$ acetic acid will be added. The mixture shall be mixed
   by shaking in an end-over-end shaker as described above for 16 h (*e.g.*
   overnight) in a room at $20 \pm 2$°C.
3. A subsample of 10 g soil shall be transferred to an extraction bottle in which
   20 mL of 0.005 mol $L^{-1}$ DTPA solution will be added. The mixture shall be
   mixed by shaking in an end-over-end shaker as described above for 2 h in a
   room at $20 \pm 2$ °C.

The temperature of the room shall be measured at the beginning and at the
end of the extraction as well as the temperature of the extracting solution in
the bottle at the end of the shaking period. The extracts shall be separated
immediately. To separate, decant a portion of the extract into a centrifuge
tube and centrifuge for 10 min at about 3000 g. The supernatant liquid must
be stored in a polyethylene container at 4°C until analysis. Alternatively, the
separation can be performed by filtration through a filter paper (porosity 0.2
to 1.1 $\mu$m capable of retaining particles of 2.7 $\mu$m size) previously rinsed with
0.05 mol $L^{-1}$ EDTA followed by distilled water. The filtrates shall be collected
in polyethylene bottles and stored at 4°C until analysis. Blank extractions (*i.e.*
without soil) shall be carried out for each set of analysis using the same
reagents as described above.

   The sample for analysis should be taken as it is. Before a bottle is opened it
should be manually shaken for 5 min to rehomogenise the content. The results
should be corrected for dry mass: this correction must be performed on a
separate portion of 1 g taken at the same time from the same bottle by drying
in an oven at $105 \pm 2$ °C for 2–3 h until constant mass is attained (successive
weightings should not differ by more than 1 mg).

# $NH_4NO_3$, $CaCl_2$ and $NaNO_3$ Extraction Protocols

In addition to EDTA, DTPA and acetic acid, weak extraction procedures were
applied by some laboratories in the course of the certification, *e.g.* 0.1 mol $L^{-1}$
$NH_4NO_3$, 0.01 mol $L^{-1}$ $CaCl_2$ and 0.1 mol $L^{-1}$ $NaNO_3$. These procedures are
outlined below. It is emphasised that they have not been validated by an

interlaboratory trial and that the values reported are given as indicative only.

# Extraction with 0.1 mol $L^{-1}$ $NH_4NO_3$

## Materials and Chemicals

1. Cleaning: all glassware and containers are cleaned with 4 mol $L^{-1}$ $HNO_3$ and rinsed with distilled water.
2. Apparatus: End-over-end shaker, centrifuge, acid-washed filter paper, 0.45 $\mu$m membrane filter, polyethylene or PTFE extraction bottles (100–150 mL) preconditioned with 4 mol $L^{-1}$ $HNO_3$ and 50 mL polyethylene bottles preconditioned with 4 mol $L^{-1}$ $HNO_3$.
3. Reagents: conc. $HNO_3$ (suprapur), 1.40–1.43 kg $L^{-1}$, 1% (v/v) $HNO_3$ and 1 mol $L^{-1}$ $NH_4NO_3$ (dissolve 80.04 g $NH_4NO_3$ in doubly distilled water).

## Procedure

Manually shake the sample bottle for 1 min to homogenise contents and take sample for analysis directly from the bottle. Operations should be carried out at $20 \pm 2\,°C$. Weigh out accurately 20.0 g soil and extract in extraction bottle on shaker operating at 50–60 rpm for 2 hours; filter supernatant solution through filter paper into 50 mL bottle, discarding the first 5 mL of filtrate; stabilise the solution by adding 1 mL $HNO_3$. If solids remain, centrifuge or filter through membrane filter. Analyse solution immediately. Carry out two blank extractions (without soil) with each set of analyses using the above procedure. Correct the results to dry mass basis by drying a separate 1 g portion (taken at the same time as the analysis sample) at $105 \pm 2\,°C$ for 2–3 hours to constant ($\pm$ 1 mg) mass.

# Extraction with 0.01 mol $L^{-1}$ $CaCl_2$

## Materials and Chemicals

1. Cleaning: all glassware and containers are cleaned with 4 mol $L^{-1}$ $HNO_3$ and rinsed with distilled water.
2. Apparatus: centrifuge, polyethylene or PTFE bottles (250 mL), end-over-end shaker.
3. Reagents: 0.01 mol $L^{-1}$ $CaCl_2$ (dissolve 1.47 g $CaCl_2.2H_2O$ in doubly distilled water or equivalent). Verify that the Ca concentration is $400 \pm 10$ mg $L^{-1}$ by *e.g.* EDTA titration.

## Procedure

Manually shake the sample bottle for 1 min to homogenise contents and take sample for analysis directly from the bottle. Operations should be carried out at $20 \pm 2\,°C$. Weigh 10.0 g soil into a 250 mL polyethylene bottle. Add 100 mL of 0.01 mol $L^{-1}$ $CaCl_2$ solution and extract on shaker for 3 hours at 30 rpm; decant about 60 mL into a centrifuge tube and centrifuge for 10 min at 3000 g. Measure and report the room temperature before and after the extraction and also the temperature of the extracting solution at the end of the extraction; measure pH in the extract before centrifugation; analyse immediately. Carry out two blank extractions (without soil) with each set of analyses using the above procedure. Filtration is not recommended because of contamination risk. Dilutions, if required, are made with acidified ($HNO_3$) $CaCl_2$ solution. Correct the results to dry mass basis by drying a separate 1 g portion (taken at the same time as the analysis sample) at $105 \pm 2$ °C for 2–3 hours to constant ($\pm 1$ mg) mass.

## Extraction with 0.1 mol $L^{-1}$ $NaNO_3$

### Materials and Chemicals

1. Cleaning: all glassware and containers are cleaned with 1 mol $L^{-1}$ $HNO_3$.
2. Apparatus: end-over-end shaker, centrifuge, 0.45 $\mu$m membrane filter, polyethylene extraction bottles (200 mL), polyethylene bottles (50 mL).
3. Reagents: 1 mol $L^{-1}$ $HNO_3$ (prepared by dilution of 70 mL concentrated suprapur $HNO_3$ (65%) with milli-Q water or equivalent), 0.1 mol $L^{-1}$ $NaNO_3$ (dissolve 8.5 g $NaNO_3$ in 1 L milli-Q water).

### Procedure

Manually shake the sample bottle for 1 min to homogenise contents and take sample for analysis directly from the bottle. Weigh accurately 40 g soil and extract in extraction bottle with 100 mL of 0.1 mol $L^{-1}$ $NaNO_3$ solution at $20 \pm 2$ °C for 2 hours at 120 rpm. Centrifuge for 10 min at 4000 g; remove the supernatant by syringe; fit the membrane filter and filter into 50 mL bottle; add 2 mL conc. $HNO_3$ to 50 mL volumetric flask and make up to volume with filtered extract (prevention of microbial growth); analyse immediately (note that solutions are stable for 1 week at $20 \pm 5$ °C).

Correct the results to dry mass basis by drying a separate 1 g portion (taken at the same time as the analysis sample) at $105 \pm 2$ °C for 2–3 hours to constant ($\pm 1$ mg) mass. Correct also the results for the dilution of the extract with $HNO_3$ (48 mL to 50 mL).

## BCR Sequential Extraction Procedure (modified)

### Apparatus

All laboratory-ware shall be of borosilicate glass, polypropylene or PTFE. Vessels in contact with samples or reagents shall be cleaned by soaking in 4 mol $L^{-1}$ $HNO_3$ (overnight) and rinsed repeatedly with distilled water before use. Perform the extractions using a mechanical end-over-end shaker, at a speed of $30 \pm 10$ rpm. Record the speed. Carry out the centrifugation at 3000 g for 20 minutes.

# Reagents

All reagents should be of analytical grade or better.

## Water

Glass-distilled water should be used throughout. Alternatively, doubly deionised and filtered water (*e.g.* MilliQ or equivalent) may be used. Simple deionised water should not be used since it may contain organically complexed metal ions.

## Solution A (Acetic Acid, 0.11 mol L$^{-1}$)

Add, in a fume cupboard, $25 \pm 0.2$ mL of glacial acetic acid to about 0.5 L of distilled water in a 1 L graduated polypropylene or polyethylene bottle and make up to 1L with distilled water. Take 250 mL of this solution (acetic acid, 0.43 mol L$^{-1}$) and dilute to 1 L with distilled water to obtain an acetic acid solution of 0.11 mol L$^{-1}$.

## Solution B (Hydroxylammonium Chloride (Hydroxylamine Hydrochloride), 0.5 mol L$^{-1}$)

Dissolve 34.75 g of hydroxylammonium chloride in 400 mL distilled water. Transfer the solution to a 1 L volumetric flask, and add, by means of a volumetric pipette, 25 mL of 2 mol L$^{-1}$ HNO$_3$ (prepared by weighing from a suitable concentrated solution). Make up to 1 L with distilled water. Prepare this solution on the same day the extraction is carried out.

## Solution C (Hydrogen Peroxide, 300 mg g$^{-1}$, i.e. 8.8 mol L$^{-1}$)

Use the hydrogen peroxide as supplied by the manufacturer *i.e.* acid-stabilised to pH 2–3.

## Solution D (Ammonium Acetate, 1.0 mol L$^{-1}$)

Dissolve 77.08 g of ammonium acetate in 800 mL distilled water. Adjust the pH to $2.0 \pm 0.1$ with concentrated HNO$_3$ and make up to 1L with distilled water.

## Blanks

Cd, Cr, Cu, Ni, Pb and Zn should be determined as follows:

1. *Vessel blank*: To one vessel from each batch, taken through the cleaning procedure, add 40 mL of Solution A. Analyse this blank solution along with the sample solutions from Step 1 (described below).
2. *Reagent blank*: Analyse a sample of each batch of solutions A, B, C and D.
3. *Procedural blank*: With each batch of extractions, a blank sample (*i.e.* a vessel

with no sediment) shall be carried through the complete procedure and analysed at the end of each extraction step.

## Sequential Extraction Procedure

Determine the extractable contents of Cd, Cr, Cu, Ni, Pb and Zn using the procedure described below. Carry out all the extractions on the sediment as received in the glass bottle. Before subsampling the sediment, shake the contents of the bottle manually for three minutes. Take the sample using a suitable plastic (see Apparatus, above) spatula.

For each batch of extractions, dry a separate 1 g sample of the sediment in a layer of about 1 mm depth in an oven ($105 \pm 2$°C) until constant weight. From this, a correction to dry mass is obtained which shall be applied to all analytical values reported (*i.e.* results shall be quoted as quantity of metal per g dry sediment).

Perform the extractions by shaking in a mechanical, end-over-end shaker at a speed of $30 \pm 10$ rpm and a room temperature of $22 \pm 5$°C. measure and report the temperature of the room at the start and at the end of each step of the extraction procedure.

Perform the sequential extraction according to the steps described below:

### *Step 1*

Add 40 mL of solution A to 1 g sediment in a 100 mL centrifuge tube, stopper and extract by shaking for 16 h at $22 \pm 5$°C (overnight). No delay should occur between the addition of the extractant solution and the beginning of the shaking. Separate the extract from the solid residue by centrifugation at 3000 g for 20 minutes and decantation of the supernatant liquid into a polyethylene container. Stopper the container and analyse the extract immediately, or store in a refrigerator at about 4°C prior to analysis. Wash the residue by adding 20 mL distilled water, shaking for 15 minutes on the end-over-end shaker and centrifuging for 20 minutes at 3000 g. Decant the supernatant and discard, taking care not to discard any of the solid residue.

### *Step 2*

Add 40 mL of a freshly prepared Solution B to the residue from Step 1 in the centrifuge tube. Resuspend by manual shaking, stopper and then extract by mechanical shaking for 16 h at $22 \pm 5$°C (overnight). No delay should occur between the addition of the extractant solution and the beginning of the shaking. Separate the extract from the solid residue by centrifugation and decantation as in Step 1. Retain the extract in a stoppered polyethylene container, as before, for analysis. Wash the residue by adding 20 mL distilled water, shaking for 15 minutes on the end-over-end shaker and centrifuging for 20 minutes at 3000 g. Decant the supernatant and discard, taking care not to discard any of the solid residue.

## Step 3

Add carefully, in small aliquots to avoid losses due to violent reaction, 10 mL of Solution C to the residue in the centrifuge tube. Cover the vessel loosely with its cap and digest at room temperature for 1 h with occasional manual shaking. Continue the digestion for 1 h at $85 \pm 2°C$, with occasional manual shaking for the first $\frac{1}{2}$ hour; in a water bath, and then reduce the volume to less than 3 mL by further heating of the uncovered tube. Add a further aliquot of 10 mL of Solution C. Heat the covered vessel again to $85 \pm 2°C$ and digest for 1 h, with occasional manual shaking for the first $\frac{1}{2}$ hour. Remove the cover and reduce the volume of liquid to about 1 mL. Do no take to complete dryness. Add 50 mL of Solution D to the cool moist residue and shake shaking for 16 h at $22 \pm 5°C$ (overnight). No delay should occur between the addition of the extractant solution and the beginning of the shaking. Separate the extract from the solid residue by centrifugation and decantation as in Step 1. Stopper and retain as before for analysis.

## Recommendations

- The calibrant solutions should be made up with the appropriate extracting solutions.
- Where ETAAS is the technique used for quantification, the method of standard additions is strongly recommended for calibration.
- As an internal check on the procedure, it is recommended that the residue from Step 3 be digested in *aqua regia* and the total amount of metal extracted (*i.e.* sum of Step 1 + Step 2 + Step 3 + Residue) compared with that obtained by *aqua regia* digestion of a separate, 1 g sample of the sediment. The residue from Step 3 should be transferred to a suitable (see apparatus) digestion vessel with about 3 mL water and should be digested following the reflux procedure (ISO Norm 11466, see attached procedure). The same procedure should be used for *aqua regia* digestion of the original sediment.

## *Aqua Regia* Extraction Protocol

The following digestion method, according to the ISO Norm 11466, was adopted as the common method for the certification campaign. This international standard has been proposed for the determination of extractable metals in soils and similar materials, and containing less than about 20% (m/m) organic carbon according to ISO 10694. Materials containing more than about 20% (m/m) organic carbon will require treatment with additional nitric acid.

## Reagents

The reagents used shall meet the purity requirements of the subsequent analysis. Their purity shall be verified by performing a blank test.

## Water

The water used shall comply with grade 2 of ISO 3696 or better. Deionised water may be used, providing that it meets the requirements given above. It is recommended that the same batch of water is used throughout a given batch of determinations and the blank determinations are carried out.

## Hydrochloric acid

$c(HCl) = 12.0$ mol $L^{-1}$, $\rho = 1.19$ g mL$^{-1}$.

## Nitric Acid

$c(HNO_3) = 15.8$ mol $L^{-1}$, $\rho = 1.42$ g mL$^{-1}$.

## Nitric acid

$c(HNO_3) = 0.5$ mol $L^{-1}$, dilute 32 mL of 15.8 mol $L^{-1}$ nitric acid with water to 1 L.

## Apparatus

Clean all the glassware by carefully immersing in warm 0.5 mol $L^{-1}$ nitric acid for a minimum of 6 h and then rinse with water.

## Reaction Vessel

Nominal volume 250 mL (round bottom flask type).

## Reflux Condenser

Straight-through type, with conical ground-glass joints. Water-cooled condensers with a minimum effective length of at least 200 mm have been found suitable. The effective length is the internal surface which is in contact with the cooling water. The overall external length of such condensers is usually at least 365 mm.

## Absorption Vessel

Non-return type. The absorption vessel is only necessary when mercury is to be determined.

## Roughened Glass Beads

Diameter 2 mm to 3 mm (or anti-bumping granules).

*Temperature-controlled Heating Apparatus*

Capable of heating the contents of the reaction vessel to reflux temperature.

## Extraction Procedure

1. Weigh approximately 3 g (see Note 1 below), to the nearest 0.001 g, of the air-dried material into the reaction vessel. (The water content of the air-dried material should be determined in a different subsample according to ISO 11465).
2. Add 0.5–1.0 mL of water to obtain a slurry, and add, while mixing, 21 mL of 12.0 mol L$^{-1}$ HCl followed by 7 mL of 15.8 mol L$^{-1}$ HNO$_3$, drop by drop if necessary, to reduce foaming.
3. Add 15 mL of 0.5 mol L$^{-1}$ HNO$_3$ to the adsorption vessel, connect the vessel to the reflux condenser and place both on top of the reaction flask.
4. Allow to stand for 16 h at room temperature to allow for slow oxidation of the organic matter of the soil.
5. Raise the temperature of the reaction mixture slowly until reflux conditions are reached and maintain for 2 h, ensuring that the condensation zone is lower than 1/3 of the height of the condenser.
6. Allow to cool slowly to room temperature.
7. Add the content of the absorption vessel, through the condenser tube, into the reaction vessel and rinse both with 10 mL of 0.5 mol L$^{-1}$ HNO$_3$.
8. Filter (cellulose-based membrane filter with a medium pore size of 8 $\mu$m) or centrifuge the extract to remove particulates (silicates and other insoluble materials), collecting the filtrate in a 100 mL graduated flask.
9. Allow all the initial filtrate to pass through the filter paper then wash the insoluble residue onto the filter paper with the minimum of 0.5 mol L$^{-1}$ HNO$_3$.
10. Fill the graduated flask with bidistilled water or 0.5 mol L$^{-1}$ HNO$_3$ up to the mark (see Note 2), close with stopper and shake. The extract thus prepared is ready for the determination of trace elements by an appropriate method.

# Notes

1. When analysing the residue from Step 3 of the BCR sequential extraction protocol, the amount of acid to attack the sample should be reduced in order to maintain the same volume/mass ratio as in the above protocol (*i.e.* 7.0 mL of 12.0 mol L$^{-1}$ HCl followed by 2.3 mL of 15.8 mol L$^{-1}$ HNO$_3$).
2. The flask containing the extract could require the addition of releasing agents, depending on the element(s) of interest and the spectroscopic method chosen. The flask should not, therefore, be filled to the mark at this stage, until the further steps in the analysis have been decided upon.

# CHAPTER 4

# Sequential Extraction Procedures for the Characterisation of the Fractionation of Elements in Industrially-contaminated Soils

C. GLEYZES, S. TELLIER AND M. ASTRUC

Université de Pau, Laboratoire de Chimie Analytique, Bio-Inorganique et Environnement, Pau, France

## 4.1   Introduction

Soils have been and are still receiving numerous diffuse and localised anthropogenic contaminations, the main sources being urban centres and industrial, agricultural or mining activities. Deposited from the atmosphere or added with fertilisers, residues or solid or liquid wastes, metals and metalloids can be immobilised in the soil or travel towards surroundings waters, plants, animals and humans.

In the soils these pollutants can be mobilised *via* the soil solution or retained by different mechanisms: precipitated with primary minerals (silicates) or secondary minerals (oxides, carbonates) or complexed by solid-state organic ligands, clays and metallic oxyhydroxides. In some cases of very high pollution such as those encountered in old industrial or mining sites, they can also be found as individual solid-state minerals[1] such as HgS, CdS, FeAsS, $CdCO_3$ *etc.*

The major causes for trace element mobilisation from the solid to the liquid phase due to modifications of the local environment are the following:[1,2]

- Acidification: a decrease in the soil solution pH induces the mobilisation of most elements through cationic exchange, hydrolysis and dissolution of some of the major chemicals such as carbonates; alkalinisation may lead to anion exchange and dissolution of metalloids;
- Change in the redox conditions: very oxidising conditions may induce dissolution of metal sulfides and destruction of organic matter; strongly reducing conditions may lead to metal sulfide precipitation following reduction of

sulfates but also to the dissolution of iron and manganese oxyhydroxides and the release of associated pollutants;

- Increase of dissolved organic ligand concentration: mobilisation of elements by complex formation and desorption;
- Increase of the concentration of inorganic salts: several effects may be observed such as ion exchange, competition for sorption reactions or the formation of soluble or insoluble complexes;
- Methylation: inorganic derivatives of some elements such as arsenic, mercury, tin or selenium can be converted by a great variety of microorganisms into methylated species that may then be accumulated in organic matter or plants or else volatilised to the atmosphere.

Several of these processes can occur simultaneously or in sequence and it is the alternation of these different conditions that causes pollutant redistribution and favours the contamination of surrounding waters. So, in order to evaluate the potential risks induced by the pollution of a soil by toxic elements, it is very important to well define not only the total concentration of the pollutant but also the predictable impact of environmental changes (pH, potential *etc.*) on its mobility.

Existing solid-state analytical methods allowing a non-destructive direct differentiation of the physico-chemical species of a trace element in solid matrices do not have a sensitivity sufficient to deal with most actual situations. Their use at this time is therefore limited to the study of very heavily polluted sites and so will not be dealt with in this paper.

In practice, the characterisation of the repartition of the pollutant in functionally defined metal fractions ('fractionation' following the recent IUPAC definition)[3] is based on the application of single, parallel or sequential selective chemical extractions and, despite numerous criticisms (lack of uniformity in the protocols, lack of selectivity, results highly dependant on the procedure used *etc.*), these schemes are frequently used.

However most of these procedures were established to deal with the case of sediments or agricultural soils and their application to industrially-polluted soils, which often have very different matrix compositions and properties, is questionable.

The aim of this work was first to evaluate the validity of the fractionation results obtained by applications of classical extraction schemes to industrially-polluted soils and then to establish reliable and easy-to-use procedures suitable for the evaluation of the potential mobility of trace elements in some of these situations.

## 4.2 Literature Search

### 4.2.1 Definitions and Problems

Trace elements occur naturally in soils as minor minerals that originate from the parent (or rock) materials as a consequence of different long-term dissolution/ immobilisation processes. They are finally redistributed at very low concentrations in waters and soils. Some of them are essential for living organisms at trace level (micro-nutrients), but become toxic beyond a given concentration. Others, such as Cd or Pb, are considered as non-essential and are only potentially toxic.

'Contamination' can be defined as the increase of the concentration of an element in the natural environment due to human activities.[4] The term 'pollution' is employed when these anthropogenic supplies have negative and significant effects on the environment.[5]

It is also useful to differentiate the terms 'polluted soils' and 'industrially-polluted soils'. Polluted soils are weathered minerals naturally enriched by other compounds such as decaying and mineralised organic matter and submitted to a more or less diffuse contamination by toxic elements. Polluted sites are industrial or mining areas, abandoned or still active, or old rubbish dumps where industrially-polluted soils (IPS) are found, as a result of the accumulation, the deposition, the burying and often the mixing of various waste products or dangerous materials together or with underlying natural soils. Their physico-chemical characteristics can be totally different from those of a classical soil. The mixing of by-products with the soil can lead to extreme values of pH, or to unusually high or low organic matter or carbonate content. On these sites the toxic elements are often not found at trace level but at high concentrations as more or less specific chemical species (Table 1).

The main problem encountered with anthropogenic trace elements is that they are accumulated in the upper layers of soils, layers where interactions with plants and mankind are maximal. They are often present under rather reactive chemical forms, whereas natural trace elements are immobilised in low mobility compounds, so the potential risks of their mobilisation and redistribution in the environment are higher.

It is necessary to consider the fractionation of the contaminant, *i.e.* to charac-

**Table 1**  *Background and excessive levels of some trace elements in surface soils*

| Element | Background | Industrially-contaminated soils | | |
| --- | --- | --- | --- | --- |
| | | Encountered values | Source | References |
| Cu | 45–70 mg kg$^{-1}$ | 400–2800 mg kg$^{-1}$ | Wood preservatives | 6,7,8 |
| | | 88–28 800 mg kg$^{-1}$ | Mining activities | 9,10 |
| | | 180–10 500 mg kg$^{-1}$ | Industrial activities | 11 |
| Pb | 1–30 mg kg$^{-1}$ | 250–142 000 mg kg$^{-1}$ | Industrial activities | 11,12,13,14,15 |
| | | 268–6880 mg kg$^{-1}$ | Mining activities | 16 |
| Zn | 70–132 mg kg$^{-1}$ | 970–50 200 mg kg$^{-1}$ | Industrial activities | 11 |
| | | 4580 mg kg$^{-1}$ | Mining activities | 10 |
| Cd | < 1 mg kg$^{-1}$ | 41–62 mg kg$^{-1}$ | Batteries | 17 |
| | | 0.2–102 mg kg$^{-1}$ | Industrial activities | 18 |
| | | 374 mg kg$^{-1}$ | Mining activities | 10 |
| As | 2 mg kg$^{-1}$ | 966–1300 mg kg$^{-1}$ | Wood preservatives | 7,19,8 |
| | | 265–14 000 mg kg$^{-1}$ | Industrial activities | 20,21,22 |
| | | 174–30 000 mg kg$^{-1}$ | Mining activities | 23,24,25,10 |

terise its association to different soil fractions or its chemical state (ionic, complexed, specifically adsorbed, *etc.*)[26] to understand its environmental behaviour. From its fractionation will come its mobility, which represents the ability of an element to move from one point to another or to change its chemical forms[27] and therefore the potential risks it generates.

## 4.2.2 Methodologies

### 4.2.2.1 Case of Elements Giving Cationic Species

Sequential selective extraction techniques are commonly used to fractionate the solid phase forms of metals in soils. Many sequential extraction procedures have been developed, particularly for sediments or agricultural soils[28-30] and extensive reviews have been published.[31-33] The Community Bureau of Reference (BCR, now Standards, Measurements and Testing programme) has launched a programme aiming to harmonise single and sequential extraction schemes for the determination of extractable trace metals in soil and sediment,[30,34] which is extensively described in Chapters 2 and 3 of this book. All these methods are based on the rational use of a series of more or less selective reagents chosen to solubilise successively the different mineralogical fractions of soils. Taking for example the well known Tessier scheme, the fractions individualised are schematically called 'exchangeable', 'carbonate bound', 'reducible', 'organically bound' and 'residual' fractions. The most often used procedures are presented in Chapter 2.

### 4.2.2.2 Case of Elements Giving Anionic Species

The methods described above were elaborated to evaluate the behaviour of cationic species. They were afterwards also applied to the fractionation of elements giving anionic species, such as arsenic.[35,36]

However, the feasibility of applying these sequential extraction schemes to the study of soil fractionation of elements such as As or Se has been questioned. Indeed these elements can exist under different oxidation states, each one having a particular behaviour. As some of the extracting reagents possess reducing or oxidising properties, they can induce a change of the oxidation state of the pollutant and therefore modify extraction results.[37]

As arsenic pollution is a major problem in a great number of polluted sites, research concentrated on this point. An attractive alternative to the use of the schemes mentioned above is to adapt soil phosphorus fractionation procedures because As and P chemistries in soils are rather similar: they both form oxyanions when they are present under their oxidised states.[38] These fractionation procedures are based on the Chang and Jackson scheme[39] (1957) which has already been used more than twenty years ago for the study of As distribution in contaminated soils[40] (Table 2).

Most of these schemes partitioned As into the following fractions: 'soluble or exchangeable', 'Al-associated', 'Fe-associated', 'Ca-associated' and 'residual', this

**Table 2** *Sequential extraction schemes for the fractionation of elements with anionic species*

| Extractants | Fraction | Element | References |
|---|---|---|---|
| $NH_4Cl$ 1 mol $L^{-1}$<br>$NH_4F$ 0.5 mol $L^{-1}$<br>NaOH 0.1 mol $L^{-1}$<br>$H_2SO_4$ 2 mol $L^{-1}$<br>CBD [a]<br>$NH_4F$ 0.5 mol $L^{-1}$<br>  and/or NaOH 0.1 mol $L^{-1}$ | soluble<br>Al-phosphate<br>Fe-phosphate<br>Ca-phosphate<br>reductant soluble Fe-phosphate<br>occluded Al-/Fe- phosphate | P | 39 |
| NaOH 1 mol $L^{-1}$ } A<br>HCl 1 mol $L^{-1}$<br>HCl 3.5 mol $L^{-1}$  B<br>HCl 1 mol $L^{-1}$ } C<br>HCl 1 mol $L^{-1}$ | extracted in NaOH<br>extracted in HCl<br>extracted in conc.[b] HCl<br>inorganic<br>organic | P | 41 |
| $NH_4Cl$ 1 mol $L^{-1}$<br>$NH_4F$ 0.5 mol $L^{-1}$<br>NaOH 0.1 mol $L^{-1}$<br>$H_2SO_4$ 2 mol $L^{-1}$ | soluble<br>Al-associated<br>Fe-associated<br>Ca-associated | As | 40,42 |
| $H_2O$<br>$NH_4F$ 0.5 mol $L^{-1}$, pH = 8.5<br>NaOH 0.1 mol $L^{-1}$<br><br>$H_2SO_4$ 0.5 mol $L^{-1}$<br>$NH_4F$ 0.5 mol $L^{-1}$, pH = 8.5 | water soluble<br>Al-associated<br>Fe- and organic Fe-<br>  associated<br>Ca-associated<br>occluded Al-associated | As | 23,25 |
| $NH_4Cl$ 1 mol $L^{-1}$<br>NaOH 0.1 mol $L^{-1}$<br>citrate–bicarbonate 0.3 mol $L^{-1}$<br>CDB[a] 0.3 mol $L^{-1}$<br>HCl 1 mol $L^{-1}$<br>$H_2O_2$ 30% / $NH_4OAc$<br>  0.8 mol $L^{-1}$<br>acids | soluble<br>non-occluded Fe/Al oxides<br>amorphous Fe<br>reducible Fe<br>Ca-associated<br>organic matter<br><br>residual | As | 24 |
| anion exchange resin<br>$NaHCO_3$ 0.5 mol $L^{-1}$, pH = 8.5<br>NaOH 0.1 mol $L^{-1}$<br>HCl 1 mol $L^{-1}$<br>*aqua regia* | exchangeable<br>easily labile<br>sorbed on Al/Fe<br>Ca-associated<br>recalcitrant As | As | 38 |
| KCl 0.25 mol $L^{-1}$<br>$K_2HPO_4$ 0.5 mol $L^{-1}$<br>$Na_2SO_3$ 1 mol $L^{-1}$<br>NaClO<br>HCl 4 mol $L^{-1}$ | soluble<br>adsorbed<br>elemental<br>organic material<br>oxides | Se | 43 |

[a] citrate–dithionite–bicarbonate
[b] concentrated HCl

last fraction representing in general As associated to the lattice of refractory crystalline minerals such as silicates.

The main difference of these schemes with classical procedures is to evaluate the 'Fe- and Al-associated' fractions of As with a NaOH extraction, after the 'exchangeable' fraction. Indeed, the solubilisation of anionic species such as Cr(VI) (*i.e.* $HCrO_4^-$, $CrO_4^{2-}$) or As(V) ($H_2AsO_4^-$, $HAsO_4^{2-}$) or else As(III) ($HAsO_2$, $AsO_2^-$) is highly dependent on the pH. It is favoured in basic conditions, adsorption to most mineral surfaces being minimal at alkaline pH values.[44–47] The 'Ca-associated' fraction is evaluated by an acidic extraction following the Fe/Al step. Depending on the nature of the sample studied, some authors include an 'organic matter' step[24] or differentiate several metal oxide classes: 'non-occluded iron and aluminium' and 'amorphous and reducible'.[24,39]

### 4.2.3  Conclusion

Numerous sequential extraction schemes have been developed in order to assess the different soil fractions responsible for the immobilisation of metallic or metalloidic pollutants and evaluate the potential risks of mobilisation.

However all these procedures were established for the study of lightly 'contaminated' soils or sediments and whatever the scheme, the meaning of the results is far from being clear when these methods are applied to an industrial site soil. This is due to the very high pollutant concentrations encountered in IPS and/or to the presence of unusual 'soil' major constituents introduced by the particular activities which were developed on site. These matrix constituents may interfere with some steps of the sequential extraction procedures due to their solubility in a given reagent. Therefore, the fractionation results obtained have to be interpreted very carefully in conjunction with a good knowledge of the major soil components.

## 4.3  Experimental Study

### 4.3.1  Presentation of the Samples Studied

A study of the application of various fractionation procedures to IPS was developed in order to allow a comparison of their significance.

Five IPS samples were studied. 'Soils' A, B, C and E were received directly from different old industrial sites. The last 'soil' sample (T) came from a gold mining site still in activity. Their chemical characteristics are listed in Table 3. Table 4 summarises the mineralogical composition of the five samples. X-ray diffraction was used for this purpose but the amorphous characteristics of most iron oxides present did not allow the composition to be identified, as only crystalline forms could be evaluated. Moreover, species representing less than 5% of the samples could not be observed.

'Soils' A, B and C are industrial site soils, with high Fe contents (14 to 46%).

'Soil' A has medium gypsum and carbonate contents, and the only crystalline

**Table 3**   *Characteristics of soils A, B, C, E and T*

| Parameters | Soil A | Soil B | Soil C | Soil E | Soil T |
|---|---|---|---|---|---|
| pH | 6.5 | 4.6 | 6.7 | 7.9 | 7 |
| OM (%) | 7.7 | 0.75 | 9 | 7 | 5 |
| carbonates (%) | 4.6 | < d.l. | 0.6 | 28 | 11.6 |
| Ca (%) | 8 | 4.3 | 0.8 | 19.5 | 10 |
| Fe (%) | 15 | 46 | 24 | 2.4 | 12.2 |
| S (%) | 3 | 9 | 1.6 | 0.23 | 1.13 |
| As (mg kg$^{-1}$) | 14 000 | 750 | 300 | 120 | 30 000 |
| Pb (mg kg$^{-1}$) | 13 000 | 12 000 | 1330 | 380 | 5130 |
| Cu (mg kg$^{-1}$) | 2540 | 1990 | 1690 | ≪I.V. | 28 800 |
| Zn (mg kg$^{-1}$) | 2620 | 3250 | 6000 | 500 | 4580 |
| Cd (mg kg$^{-1}$) | 15 | ≪I.V. | 78 | ≪I.V. | 374 |
| Cr (mg kg$^{-1}$) | 55 | ≪I.V. | ≪I.V. | ≪I.V. | 604 |
| Mn (mg kg$^{-1}$) | 415 | 95 | 450 | 600 | 1794 |

OM: Organic matter
< d.l.: inferior to the detection limit
≪I.V.: inferior to the intervention values of Netherlands[48]

**Table 4**   *X-ray diffraction results*

| Soil A | Quartz +++ | Calcite ++ | Gypsum ++ | Hematite ++ | Spinel + | | |
|---|---|---|---|---|---|---|---|
| Soil B | | | Gypsum ++ | Hematite ++ | Jarosite ++ | Maghemite + | Lepidocrocite + |
| Soil C | Quartz +++ | Calcite + | | Hematite ++ | | | |
| Soil E | Quartz +++ | Calcite +++ | | | | | |
| Soil T | Quartz +++ | Calcite + | | | Dolomite ++ | Pyrite ++ | |

+ + + : high content
+ + : medium content
+ : low content

form of iron which has been evidenced is hematite. Showing a pH close to the neutrality, it is highly polluted with arsenic, lead, copper and zinc.

'Soil' B presents a very high iron content (46%) and an acidic pH. Poor in organic matter, showing a sandy texture (75% sand), it consists of different iron oxides such as hematite, lepidocrocite, maghemite and more particularly jarosite. It is highly polluted with lead, and to a lesser extent with arsenic, copper and zinc.

'Soil' C mineralogy appears similar that of 'soil' A. Consisting of quartz, calcite and hematite, its pH is also close to neutrality and its organic matter content is quite high (9%). It is polluted with arsenic, cadmium, copper, lead and zinc.

'Soil' E originates from an industrial waste pond sludge; it presents a high limestone content (28%) and a slightly alkaline pH (7.9). Its constituents are essentially quartz and calcite and no detectable concentration of crystalline iron oxide was found. The better part of iron oxides should then be essentially amorphous. In contrast to the other samples, pollutant concentrations in this soil are inferior to or close to the limits of intervention enforced in The Netherlands.[48]

'Soil' T is a mining site soil with a neutral pH, highly polluted with arsenic, copper and cadmium, and to a lesser degree with lead, zinc and chromium. The major crystalline constituents are quartz and dolomite with a low content of calcite and pyrite.

## 4.3.2   Application of Classical Sequential Extraction Schemes

### 4.3.2.1   Study of Elements Giving Cationic Species: Lead and Copper

Two different sequential extraction schemes were applied to the five soil samples. A description of these two procedures is given in Appendix 2.

The first one was the classical Tessier scheme.[28] It fractionates the soil sample in five fractions: exchangeable (TF1), bound to carbonates (TF2), bound to manganese and iron oxides (TF3), bound to organic matter (TF4) and residual (TF5).

The second, called Scheme 1, previously established by Delmas,[49] was based on the Tessier scheme. Due to the high iron contents of the studied samples the reduction step was subdivided into three fractions according to the Shuman scheme[29] because it has already been shown that the use of hydroxylamine in the conditions of the Tessier scheme (step TF3) was not sufficient, particularly when the iron content of the sample was high.[50] Although this subdivision lengthens the experimental protocol, it allows a better understanding of the role played by iron oxides in pollutant scavenging.

The fraction related to amorphous iron oxides was extracted in the dark with an oxalate/oxalic acid solution. Crystalline iron oxides were then extracted with the same reagent associated to ascorbic acid. Seven fractions were thus obtained with the so-called Scheme 1: exchangeable (MTF1), bound to carbonates (MTF2), bound to manganese oxides (MTF3), bound to amorphous iron oxides (MTF3a), bound to crystalline iron oxides (MTF3c), bound to organic matter (MTF4) and residual (MTF5).

Results obtained from the application of these two schemes to the five IPS samples in the case of Pb and Cu fractionation are presented in Tables 5, 6, 7 and 8.

### Application of the Tessier Scheme

**Lead**

It appears that lead is essentially extracted in the residual fraction (more than 85%) with the Tessier scheme, except in the case of soil E (24%) (Table 5).

**Table 5**   *Pb fractionation according to the Tessier scheme*

|        | TF1    | TF2    | TF3   | TF4     | TF5    |
|--------|--------|--------|-------|---------|--------|
| Soil A | 0.8%   | 6%     | 6%    | 1.1%    | 85%    |
| Soil B | 1.2%   | 1.9%   | 1.2%  | 4%      | 92%    |
| Soil C | 0.06%  | 3%     | 5.4%  | < d.l.  | 91.5%  |
| Soil E | 0.4%   | 25.5%  | 44%   | 6.7%    | 24%    |
| Soil T | 0.06%  | 2.8%   | 4.1%  | 1.3%    | 92%    |

The exchangeable fraction represents only 1% of the total concentration of lead in the case of soils A and B and is negligible in the others. These results are in agreement with the literature. Lead is generally not found in an exchangeable form, except in some cases of important specific industrial pollution.[13,14] Lead associated to the carbonate fraction is inferior to 3% for soils B, C and T. For soil A it represents 6%. This soil shows a low carbonate content; it is possible that Pb extracted was initially weakly sorbed on iron oxides or weakly associated with organic matter and that a small part of it was released during the carbonate attack step. This extraction yield reaches 25.5% for soil E; this is in agreement with its high carbonate content.

The 'associated to iron and manganese oxides' (or 'reducible') fraction of lead (TF3) is lower than 10%, except for soil E where it amounts to 44%. These results are quite surprising because the studied IPS have a high iron content and lead is generally described as presenting a strong affinity for iron oxides.[51-53]

During the oxidation step, less than 10% of total lead is extracted. According to the literature, this fraction plays an important role in Pb retention only in the case of a very high organic matter content.[54] This amount being inferior to 10% for all the studied IPS, these results seem reliable.

## Copper

In the case of copper, a high contribution of the residual fraction is also obtained (between 58 and 86%), particularly in the case of soils B and C (Table 6).

However it may be noted that for soil T, 38% of Cu is extracted during the second and third steps (19.7 and 18.3%). According to Waller *et al.*,[25] in the case of mining wastes having a low organic matter content, copper is essentially extracted in the acid-soluble extraction. To understand these figures it is also important to notice that this IPS has an extremely high Cu content (28 800 mg kg$^{-1}$ *i.e.* 0.45 mol kg$^{-1}$). It is possible that the quantities of reagents employed in the fractionation scheme during this second step are not sufficient to extract all

**Table 6**   *Cu fractionation according to the Tessier scheme*

|        | TF1    | TF2    | TF3   | TF4    | TF5    |
|--------|--------|--------|-------|--------|--------|
| Soil A | 1.8%   | 12%    | 0.9%  | 25.6%  | 60%    |
| Soil B | 0.8%   | 2%     | 7.4%  | 8.6%   | 81%    |
| Soil C | 0.04%  | 2%     | 5.9%  | 6.3%   | 86%    |
| Soil T | 0.03%  | 19.7%  | 18.3% | 3.5%   | 58.5%  |

the copper of this fraction or that re-precipitation due to chemical saturation occurs. Therefore the Cu extracted in the following step (TF3) may belong to the preceeding fraction and not be actually associated to manganese or iron oxides.

## Application of Scheme 1

This scheme uses the same reagents as the Tessier scheme except for the addition of two supplementary steps specific for the various kinds of iron oxides. So, in this modified scheme, step 'MTF3' (identical to the 'TF3' step of the Tessier scheme) is supposed to extract manganese oxides, whereas amorphous iron oxides and crystalline iron oxides should be dissolved respectively during the 'MTF3a' and 'MTF3c' steps.

The 'MTF3a' reagent is a buffered oxalate/oxalic acid solution at pH = 2; this is a weakly reducing reagent with complexing properties towards iron. The following 'MTF3c' step uses the effects of the oxalate solution combined with a more energetic reducing reagent, ascorbic acid.

A comparison between the final redox potentials of solutions after IPS extraction with hydroxylamine and oxalate is presented in Table 7 using the notion of pe + pH to define the redox conditions of the system.[55] The scale varies between 0 (reducing) and 21.6 (oxidising). Values of pe + pH ranging between 10 and 18 represent oxidising, 7 to 9 moderately reducing and pe + pH < 7 reducing conditions.

The results obtained in this study show that the step corresponding to iron and manganese oxide dissolution in the Tessier scheme (called 'MTF3' in Scheme 1) leads to pe + pH values ranging between 9.5 and 11, which are not reducing conditions. On the other hand, the application to IPS samples of oxalate and ascorbic acid/oxalate solutions (steps MTF3a and MTF3c) allows to decrease the final potential (pe + pH ranging between 5.5 and 8.5) and favours the solubilisation of iron oxides (Table 8).

So, the use of hydroxylamine as described in the Tessier scheme does not allow evaluation of the role that iron plays as pollutant scavenger in IPS. It is necessary to add supplementary steps employing more specific reagents to favour iron oxides dissolution.

**Table 7**   *Final potentials of the fractions MTF3, MTF3a and MTF3c*

| Fraction | | Soil A | Soil B | Soil C | Soil E | Soil T |
|----------|--------|--------|--------|--------|--------|--------|
| MTF3 | Eh (V) | 0.506 | 0.536 | 0.531 | 0.485 | 0.397 |
| | pe + pH | 10.4 | 10.3 | 11 | 10.5 | 9.5 |
| MTF3a | Eh (V) | 0.195 | 0.261 | 0.215 | 0.365 | 0.293 |
| | pe + pH | 5.5 | 6.8 | 6 | 8.5 | 7.3 |
| MTF3c | Eh (V) | 0.159 | 0.175 | 0.162 | 0.096 | 0.293 |
| | pe + pH | 4.9 | 5.2 | 5.1 | 3.8 | 7.1 |

**Table 8**   *Iron extraction yields during the reducing steps*

| Fraction | Soil A | Soil B | Soil C | Soil E | Soil T |
|----------|--------|--------|--------|--------|--------|
| MTF3 | 3.4% | 3.2% | 5.6% | 26% | 16.5% |
| MTF3a | 19.7% | 19.5% | 11% | 13% | 20.9% |
| MTF3c | 17% | 8.5% | 18.5% | 21% | 10% |

**Table 9**   *Pb fractionation according to Scheme 1*

|  | MTF1 | MTF2 | MTF3 | MTF3a | MTF3c | MTF4 | MTF5 |
|--|------|------|------|-------|-------|------|------|
| Soil A | 0.9% | 6.2% | 7% | 4.7% | 5.3% | 36.2% | 39.7% |
| Soil B | 1% | 0.6% | 1% | 3.1% | 4.1% | 35% | 55% |
| Soil C | 0.01% | 5.2% | 4.9% | 45.4% | 21% | 4.9% | 18.6% |
| Soil E | 0.4% | 21.9% | 44.9% | 3.9% | 1.4% | 2.7% | 24.7% |
| Soil T | 0.05% | 2.1% | 2.6% | 0.9% | 7.3% | 14.5% | 72.6% |

## Lead

As regards Pb fractionation, little changes were observed for soil E (Table 9). The residual fraction is the same in the two schemes (24 and 24.7%) and the easily mobile Pb content (MTF2 + MTF3) represents about 67%. This soil has low Pb and Fe contents in comparison with the others, which can explain its different behaviour.

On the other hand significant differences are obtained for the other soils. In the case of soil C these further steps (MTF3a and MTF3c) allow the extraction of 45.4% and 21% of Pb respectively. So, on the contrary to what could be concluded from the application of the Tessier scheme, most of the lead content could be extracted under reducing conditions (pe + pH = 6) and the lead associated with the residual fraction is reduced to the order of 18%.

Soils A, B and T present higher total Pb contents (5100 to 12 800 mg kg$^{-1}$) and a similar repartition amongst the various phases. The two supplementary steps do not allow the extraction of more Pb in reducing conditions. On the other hand it is seen that the organic matter fraction is modified: 36%, 35% and 14.5% respectively of the lead content is extracted from soils A, B and T instead of 1.1, 4 and 1.3%. This result is quite surprising at first sight because the operating conditions of this extraction are identical in the two schemes. It would seem that the new reducing steps of Scheme 1 have modified lead distribution in the sample. This supposition could be confirmed by the application of a single direct extraction to the initial IPS samples to evaluate their organically bound lead fraction.[56] The results obtained in the case of soils A and B indicated that this fraction was lower than 10%, as was already found using the Tessier scheme. So the application of an oxalate extraction to IPS samples would induce a modification of the distribution of lead, particularly in the case of high pollutions (soils A and B).

The literature[57] indicates that oxalate is a ligand presenting a good affinity for $Fe^{3+}$, $Al^{3+}$ and some other metals but lead appears as one of the less strongly complexed elements:

$$Fe^{3+} > Al^{3+} \gg Cu^{2+} > Zn^{2+} > Pb^{2+}, Fe^{2+} > Cd^{2+}$$

The solubility product of lead oxalate is also very low ($2.74 \times 10^{-11}$ at $18\,^\circ$C).[58]

Oxalate can also form a precipitate with Ca; a crystalline calcium oxalate compound was effectively evidenced by applying X-ray diffraction to the solid residue after treatment of soil T. It may then be concluded that the competition for ligands between the different elements, and more particularly with iron, favours a lower Pb extraction yield. The precipitation of noticeable amounts of Ca-oxalate can also result in the scavenging of dissolved lead. Later on, during the following extraction step (MTF4), the use of an ammonium acetate solution allows the formation of Pb-acetate complexes (log $K$ ranging between 2.1 and 3.4) resulting in the leaching of lead fractions which have reacted earlier but were re-adsorbed or re-precipitated during the previous steps. This phenomenon would be the more important in the case of high lead pollution.

### Copper

The use of Scheme 1 also leads to a distribution of copper different to that found using the Tessier scheme (Table 10). The residual fraction is inferior to 35% in each case, this confirms the overestimation of this fraction with the Tessier scheme.

As in the Tessier scheme, amounts extracted during the steps 'MTF2' and 'MTF3' from soil T represents more than 60% of total copper. The extracted calcium amount seems to indicate that the dissolution of carbonates is incomplete during the devoted step 'MTF2' and continues during the step 'MTF3'. Repeated (three) successive extractions with acetate/acetic acid (step 'MTF2') show that up to 66% of Cu may be dissolved at pH = 4.5. So the better part of copper in this sample is very mobile under moderately acidic conditions.

For soils A, B and C, the step 'MTF3a' allows the extraction of respectively 36%, 50.2% and 18.4% of total copper. The following step ('MTF3c') using more reducing conditions (pe + pH = 5) allows the extraction of a further 25% of Cu from soil C. For this last sample, it seems that a noticeable proportion of total copper is incorporated in a crystalline iron oxide, the dissolution of which requires quite a strongly reducing environment. In the case of soil B, XRD analyses of the solid residues following oxalate extraction ('MTF3a') show that jarosite has been totally dissolved. It has already been demonstrated that the oxalate reagent can attack not only amorphous iron oxides but also some crystalline forms.[59-61] Accordingly, it could be concluded that Cu in soil B could

**Table 10**   *Cu fractionation according to Scheme 1*

|        | MTF1    | MTF2  | MTF3 | MTF3a | MTF3c | MTF4  | MTF5  |
|--------|---------|-------|------|-------|-------|-------|-------|
| Soil A | 3.9%    | 13.6% | 1.9% | 36%   | 1.9%  | 30.6% | 12.1% |
| Soil B | 0.9%    | 1.9%  | 4.7% | 50.2% | 1.9%  | 7.1%  | 33.2% |
| Soil C | < 0.05% | 3.3%  | 4.6% | 18.4% | 25%   | 19.2% | 29.5% |
| Soil T | 0.1%    | 40%   | 25%  | 9%    | 0.1%  | 7%    | 25%   |

be bound to amorphous iron oxides or jarosite. This supposition is further supported by Herbert[52] who suggested that Cu could co-precipitate with jarosite when this mineral is present in soils.

The amounts of Cu associated to the 'organically bound' fraction are the same in the two schemes, excepted in the case of soil C. This fraction, as for the study of lead extraction, was therefore evaluated with a direct single extraction.[56] The results show that both sequential extraction schemes overestimate the amount of Cu associated to organic matter in the case of soil B and even more for soil C. Having a well known high affinity for organic matter, it is possible that a part of the Cu solubilised from other mineralogical phases during the previous extraction steps has been retained by complexation with solid organic matter and is released in this fraction.[62]

## Conclusion

The use of the classical Tessier scheme for the evaluation of the fractionation of Cu and Pb in highly polluted industrial soils lead to a similar distribution in most cases: the metals are essentially attributed to the residual fraction.

Scheme 1 makes use of the same reagents as the Tessier scheme with the exception of two supplementary steps specific to the dissolution of iron oxides. The consideration of the pe + pH variations shows clearly that the reducing step of the Tessier scheme (TF3 or MTF3 in Scheme 1) employing hydroxylamine is not aggressive enough to dissolve iron oxides leading therefore to underestimate their contribution to the retention of cationic pollutants. On the other hand, although oxalate is specific to iron extraction, the precipitation of calcium oxalate and the low stability of complexes with lead favour a redistribution of this pollutant to other phases of the solid and Scheme 1 also does not allow a satisfying evaluation of the fraction of lead soluble in reducing conditions.

## 4.3.2.2   Study of Elements Giving Anionic Species: Arsenic

Three sequential extraction schemes were applied to the five IPS samples. The first objective was to evaluate the ability of the schemes presented above and designed to study the fractionation of metal cations (Tessier scheme and Scheme 1) to deal with the case of an important anionic pollutant: arsenic. The results were then compared to those obtained with a procedure established for P fractionation in soils (Scheme 2).

## Classical Sequential Extraction Schemes

The results obtained with the Tessier scheme and Scheme 1 are listed in Tables 11 and 12. The same general conclusion holds when the Tessier scheme is used: As appears essentially associated to the residual fraction.

Application of Scheme 1 shows that As is essentially extracted from the amorphous iron oxide fraction (step MTF3a) though the total iron content of the various IPS varied widely. In order to confirm these results, samples were also

**Table 11**   As fractionation according to the Tessier scheme

|        | TF1  | TF2   | TF3   | TF4  | TF5   |
|--------|------|-------|-------|------|-------|
| Soil A | 0.2% | 2.3%  | 8.3%  | 3%   | 86.2% |
| Soil B | 1.3% | 0.4%  | 5.4%  | 0.8% | 92%   |
| Soil C | 0.1% | 0.9%  | 7%    | 4.3% | 88%   |
| Soil E | 0.7% | 1.1%  | 17.1% | 6.8% | 74%   |
| Soil T | 0.5% | 12.1% | 12.1% | 0.3% | 75%   |

**Table 12**   As fractionation according to Scheme 1

|        | MTF1  | MTF2  | MTF3  | MTF3a | MTF3c  | MTF4     | MTF5  |
|--------|-------|-------|-------|-------|--------|----------|-------|
| Soil A | 0.02% | 2.3%  | 10.5% | 81.9% | 0.5%   | 0.01%    | 4.7%  |
| Soil B | 2%    | 0.3%  | 4.1%  | 81.2% | 2%     | < 0.01%  | 10.3% |
| Soil C | 0.3%  | 1%    | 6.8%  | 60%   | 5%     | < 0.3%   | 27.1% |
| Soil E | 5.4%  | 1.1%  | 32.2% | 34.3% | <0.5%  | 0.7%     | 25.8% |
| Soil T | 0.5%  | 14.2% | 14.2% | 56.8% | 0.1%   | 0.02%    | 14.2% |

submitted to repeated (three) extractions with the oxalate/oxalic acid reagent, operating conditions being the same as those used during sequential extractions. These results show the same tendency: As extraction and Fe dissolution seem simultaneous. Iron oxides are known to be excellent scavengers for As species especially arsenate ions.[55,63,64] Hydrous Fe oxides and hydroxides surfaces have a zero charge pH ranging from 7 to 10, with a mean value close to 8.5; lower pH values favouring a net positive charge. The pH of studied IPS being inferior to 8.6, Fe oxide surfaces are more or less positively charged and therefore suitable for the adsorption of As oxyanions from the soil solution.

Considering soil B, jarosite had been totally dissolved during this extraction. As before we can suppose that As could as well have been bound to this mineral in soil B. This supposition is strengthened by Foster et al.[65] who suggested that arsenate could be substituted for sulfate in the structure of jarosite $(KFe_3(SO_4)_2(OH)_6)$.

In the particular case of soil T, which has a high carbonate content, it may be noted that 14% of As was found in fraction MTF2. Sadiq[55] explains that in alkaline soils, carbonate minerals are expected to adsorb arsenic, and that, after the principal reaction with iron oxides, the reactive levels of Ca in calcareous soils controls the dominant forms of arsenic. Juillot et al.[66] also identified the formation of different calcium arsenates in contaminated surficial environments at an industrial site. The XRD analysis of the solid residues of three consecutive extractions with the acetate/acetic acid reagent of soil T shows that only calcite and a part of dolomite were dissolved during this extraction. The mobilisation of 25% of total As (in three extractions) from this soil could be induced by the dissolution of calcite and dolomite.

A quite similar behaviour may be observed for the soil E (containing 38% $CaCO_3$) but the results are not so significant.

A validation of the hypothesis presented above, *i.e.* iron governs the leaching of arsenic in most cases, was attempted by studying the correlation between As extracted and Fe dissolved. A good correlation was observed for soil A ($r^2 = 0.965$), soil B ($r^2 = 0.986$), soil C ($r^2 = 0.912$) and soil E ($r^2 = 0.897$). However lower results were obtained for soil T ($r^2 = 0.671$), which indicates that another factor can influence As scavenging, as an important calcium content. This sample originating from a mining site, it appears quite likely that its arsenic content originates from the weathering of arsenopyrite AsSFe into several other more easily soluble arsenic species, some of which only contain iron. From all these observations appears however a quite general trend towards iron dissolution governing As leaching.

## *Scheme 2*

Scheme 2 was originally designed to study the potential mobility of phosphorus in soils.[39]

In this scheme arsenic is fractionated into the following fractions:[23] water soluble (WS) (water), Al-associated (Al) (ammonium fluoride 0.5 mol $L^{-1}$), Fe-associated (Fe) (sodium hydroxide 0.1 mol $L^{-1}$), Ca-associated (Ca) (sulfuric acid 0.5 mol $L^{-1}$), occluded Al-associated (AlO) (ammonium fluoride 0.5 mol $L^{-1}$) and the residual fraction (R) (obtained by alkaline fusion).

This scheme was applied only to soils A and T, as well as to another IPS (soil D) previously studied in our laboratory,[49] coming from a mining site. It presents an acidic pH (5.2), a medium iron content (40 700 mg $kg^{-1}$) and is very highly polluted with arsenic (25 000 mg $kg^{-1}$). The speciation study carried out by Delmas[49] with Scheme 1 had put in evidence results similar to those presented above: the better part of As (74%) was bound to the iron oxides fraction.

The distribution of arsenic obtained with Scheme 2 (Figure 1) depends on the studied soil, except for the first fraction (water-soluble As) which is lower than 1% in each case.

In the case of soil A, the Fe-associated fraction represents only 29% of total

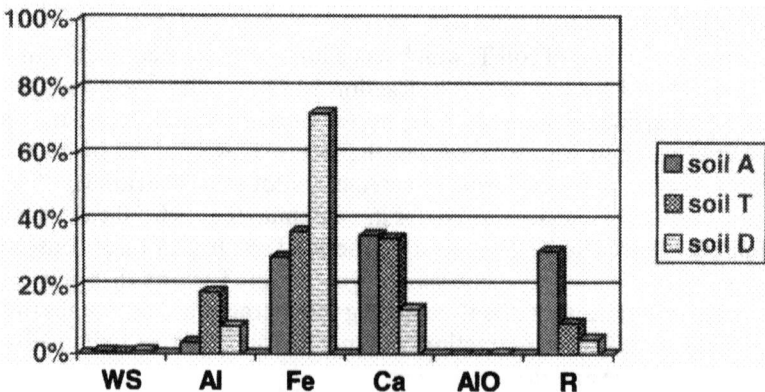

**Figure 1**  *As distribution according to Scheme 2*

As. By comparison with the results obtained with Scheme 1 (82%), it is clear that this fraction is underestimated. In Scheme 2 it is evaluated by a NaOH extraction. This reagent was originally used by Williams[67] in order to differentiate the various soil phosphorus compounds. This author showed that iron and aluminium phosphate minerals present a high solubility in soda solutions. However Chang and Jackson[39] specified that compounds soluble under reducing conditions were not extracted in soda.

On the other hand 37% and 72.5% of As are extracted during this step from soils T and D, this is not very different from the amount extracted with oxalate (respectively 57% and 74%) (step MTF3a in Scheme 1), taking into account experimental uncertainty.

These three soils present a different origin: soils T and D are mining site soils whereas soil A comes from an industrial site. Applications found in the literature of this scheme to soils presenting similar characteristics led to the same conclusions: As is extracted by NaOH when the sample originates from a mining site[23-25] whereas in the case of industrial soils As is not extracted.[23] Previous studies in our laboratory[22,49] using soda as an extracting reagent for soil decontamination led to similar conclusions.

According to these multiple observations it may be supposed that As extracted in the NaOH-fraction originates from solid compounds soluble in alkaline conditions, such as iron arsenate. Reducing conditions are necessary to extract As from industrial soils, this appears compatible with the dissolution of iron oxyhydroxides. Moreover, because of the high pH of the extracting solution, iron dissolved from iron hydroxides could have re-precipitated as a less soluble Fe–As compound.

Thus a small part of As in soil A is not extracted during this step: this leads to overestimate the following fraction (Ca), which corresponds to As associated with Ca compounds. A more detailed study of this extraction indicates that it is not fully specific, because the very acidic conditions used can dissolve not only Ca compounds but also Fe compounds. Effectively, 12%, 40% and 46% respectively of total iron were dissolved from soils A, T and D during this step. One may therefore conclude from all these observations that it is not possible with this scheme to differentiate As bound to iron species from As bound to Ca compounds at least in the case of industrial or mining site soils.

Finally, although this scheme is specific to anionic species, it is not really suitable to characterise arsenic fractionation in these iron rich soils: it underestimates the As potential mobility in reducing conditions and overestimates the potential mobility in acidic conditions.

## Conclusion

Through this study it appears that Scheme 1 is the most convenient for the evaluation of the potential of mobilisation of arsenic for the whole set of samples considered. The use of a specific iron oxide reagent – an oxalate solution – has clearly shown the role iron oxides play in arsenic scavenging and has put in evidence As mobilisation in reducing environments. In the case of a mining soil

(soil T), the existence of more soluble species, easily dissolved under acidic conditions, was also demonstrated.

### 4.3.2.3   Conclusion

Comparison of the results obtained by application of different fractionation schemes to a set of industrial and mining site soils showed that the estimations of the fractionation of the various elements considered were highly dependent on the procedure used. The classical Tessier scheme overestimated the residual fraction in all cases and gave a far too low a risk assessment. A more complex scheme (Scheme 1) using specific iron reagents allowed better evaluation of the 'reducible' fraction and underlined in the case of arsenic and copper a very good correlation between iron dissolution and pollutants leaching. Thus, with this type of soil, it appears absolutely necessary to include an extraction step characteristic of iron oxides. On the other hand, although oxalate solutions are specific to iron extraction, they induce a modification of the distribution of lead, particularly in the case of high pollution. So this reagent does not allow to evaluate correctly Pb mobilisation risk in reducing conditions.

In the case of multiple pollution, it appears thus necessary to find other reducing agents.

## 4.3.3   Comparison of Different Reducing Agents

Different reducing solutions were studied in which a complexing reagent, EDTA, was introduced because it forms stable and highly soluble complexes with most metallic cations ($Pb^{2+}$, $Cu^{2+}$, $Zn^{2+}$, $Cd^{2+}$, $Fe^{2+}$, $Fe^{3+}$, $Al^{3+}$ etc.). The mixed solution EDTA–reducing reagent will allow to maintain in solution the pollutants transferred to solution by dissolution of iron oxides.

Each extraction has been compared to a single EDTA extraction in order to differentiate the effects of the reducing reagent from those of the complexing reagent. However, although EDTA is able to solubilise amorphous iron oxides, the reaction times are very long (three weeks to several months).[68] Extraction times used in this work being much shorter, iron oxide solubilisation by EDTA alone should be minimal. In this study, realised for a part in collaboration with the Technical University of Hamburg-Harburg (Germany),[69] the behaviour of lead was more particularly examined.

The operating conditions are described in Appendix 3.

### 4.3.3.1   Extraction with TiCl₃

The first reagent tested was $TiCl_3$ which has already been used by Heron et al.[70] to evaluate the oxidative capacity of sediments. The conclusions of these authors are that the EDTA–$Ti^{3+}$ complex is able to dissolve synthetic iron oxides and oxyhydroxides such as ferrihydrite, akageneite, goethite, hematite and pyrolusite. The non-extracted iron compounds are Fe(II), very stable Fe(III) oxides and iron incorporated in silicates or clays.

In the present study the EDTA–$Ti^{3+}$ solution was used at pH$=4.6$ and two $Ti^{3+}$ concentrations were used (0.008 and 0.0292 mol $L^{-1}$) corresponding to initial redox potentials ranging between $-0.095$ V and $-0.190$ V. The EDTA concentration was 0.1 mol $L^{-1}$. In another experiment (d) the initial concentration of $Ti^{3+}$ was 0.008 mol $L^{-1}$ but solution pH and potential were maintained constant throughout the experiment by small additions of $Ti^{3+}$ and $NH_3$.

The results related to soils A and T are presented in Figures 2 and 3.

## Soil A

In the case of soil A, a single extraction with EDTA (a) leads to extraction of 55% of total lead (Figure 2). According to Elliott et al.[13], the use of EDTA in basic

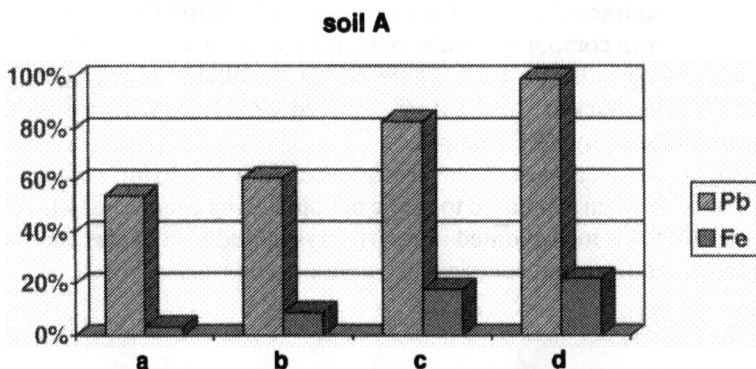

**Figure 2** *Pb and Fe extractions with EDTA-$Ti^{3+}$ solutions (soil A): (a) EDTA 0.1 mol $L^{-1}$ (Eh$=0.392$ V); (b) EDTA 0.1 mol $L^{-1}$+$TiCl_3$ 0.008 mol $L^{-1}$ (initial Eh$=-0.190$ V); (c) EDTA 0.1 mol $L^{-1}$+$TiCl_3$ 0.0292 mol $L^{-1}$ (initial Eh$=-0.095$ V); (d) EDTA 0.1 mol $L^{-1}$+$TiCl_3=f(t)$ (Eh$=-0.010$ V to Eh$=-0.110$ V)*

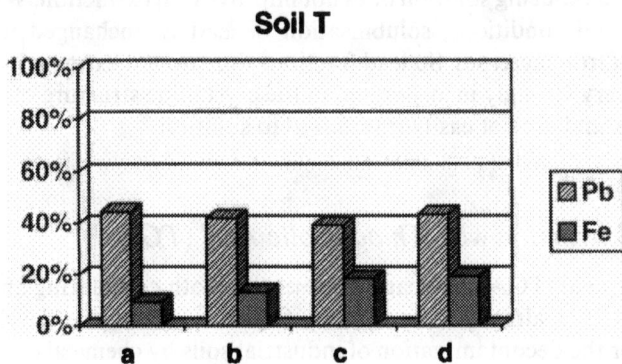

**Figure 3** *Pb and Fe extractions with EDTA-$Ti^{3+}$ solutions (soil T): (a) EDTA 0.1 mol $L^{-1}$ (Eh$=0.392$ V); (b) EDTA 0.1 mol $L^{-1}$+$TiCl_3$ 0.008 mol $L^{-1}$ (initial Eh$=-0.190$ V); (c) EDTA 0.1 mol $L^{-1}$+$TiCl_3$ 0.0292 mol $L^{-1}$ (initial Eh$=-0.095$ V); (d) EDTA 0.1 mol $L^{-1}$+$TiCl_3=f(t)$ (Eh$=-0.010$ V to Eh$=-0.110$ V)*

conditions allows to extract lead fixed by oxides and organic matter without dissolution of the solid fraction. Taking into account the pH of the extracting solution (4.6) and the complexing capacity of EDTA, this solution can also dissolve calcite and gypsum. The XRD analysis of the solid residue confirmed this hypothesis, although calcium carbonate dissolution was not complete. The percentage of total iron extracted being inferior to 2%, it may be supposed that the lead extracted was incorporated in calcite and gypsum and/or fixed on oxides and organic matter.

The addition of a reducing reagent favours lead extraction to an extent varying with $Ti^{3+}$ concentration and a quantitative extraction is obtained only with test (d) which corresponds to a solution in which the redox potential is maintained low throughout the experiment. An excellent correlation between lead extracted and iron dissolution ($r^2 = 0.97$) shows that the extraction of the 45% of total Pb which was not extracted with the 0.1 mol $L^{-1}$ EDTA solution is correlated to the dissolution of iron compounds under reducing conditions. The XRD analysis of the solid residue indicates that, whatever the conditions of extraction used, hematite is not attacked. So the iron extracted would come from amorphous oxides not detected by XRD analysis.

So lead in this soil appears to be present essentially in two forms: a large part (55%) is mobile when submitted to acidic or complexing conditions, whereas the other part (45%) is incorporated in poorly crystallised iron oxides and may be mobilised in reducing and complexing conditions (pe + pH = 3 or 4).

## Soil T

The EDTA solution extracts half of total lead present in soil T (Figure 3). According to XRD analysis of the solid residues, dolomite is not attacked, only calcite being dissolved. On the other hand, pyrite is dissolved for a part, which explains the leaching of 8.2% of iron.

The use of a reducing solution does not improve lead extraction. Whatever the redox potential conditions, solubilisation of lead is unchanged whereas the extraction of iron increases. So lead fractions that are not extracted with EDTA seem to be very strongly incorporated in the crystalline structure of some of the soil minerals and are not easily transferred to solution.

### 4.3.3.2   Extraction with Thioglycolic Acid (TGA)

Thioglycolic acid (TGA) is a reagent presenting both complexing and reducing properties. It has already been used by Segues[22] during the development of processes for the decontamination of industrial soils by chemical extraction.

Different TGA concentrations were tested (0.01 to 1 mol $L^{-1}$). Extractions realised in the presence of TGA alone are not presented here because the results have shown that the complexing capacity of TGA was not sufficient to maintain lead in solution at these high concentrations. Therefore a 0.1 mol $L^{-1}$ EDTA concentration was added to all TGA solutions. The redox potential ranged between 0 and 0.04 V (pe + pH in the order of 4). The results are summarised in Table 13.

In the case of soils A and T, results similar to those observed with TiCl$_3$ and described above were obtained.

In the case of soil B, the EDTA solution alone does not extract Pb and Fe. On the other hand iron extraction increases with TGA concentration. Simultaneously, the dissolution of iron oxides is accompanied by the total extraction of lead. So, contrary to the conclusions reached when using the Tessier scheme or Scheme 1, all the lead present in this soil appears bound to iron oxyhydroxides and is soluble in reducing conditions. Several well crystallised iron oxides were evidenced by XRD analysis of this sample, including jarosite. According to Herbert,[52] jarosite precipitation can involve Pb scavenging under the form of plumbojarosite (Pb(Fe$_3$(SO$_4$)$_2$(OH)$_6$)$_2$), Pb$^{2+}$ being able to be substituted for K$^+$ in the structure of jarosite.

Results related to soil C are in agreement with those obtained with Scheme 1: the better part of lead is soluble in reducing conditions. However 30% of total lead is extracted with EDTA alone, which represents an amount of lead easily dissolved superior to that found with Scheme 1 (6.4% during MTF2 and 6% during MTF3).

In the case of soil E, the better part of lead is extracted with EDTA and appears thus as quite easily soluble. It would seem that lead extraction from this soil is not bound to iron dissolution as it is not sensitive to a reducing environment; this is in agreement with the conclusions obtained from the application of Scheme 1.

### 4.3.3.3 Extraction with Ascorbic Acid (AA)

Ascorbic acid (AA) is a diacid also presenting reducing properties. It is used in particular in association with oxalate to attack crystalline iron oxides.[71] The reaction mechanism has been studied: the reagent is adsorbed onto the crystalline surface of iron oxides to form a surface complex with Fe(III). The dissolution rate depends on the surface concentration of ascorbate and increases when the pH decreases. There is then an electron transfer followed by the release of Fe(II) in solution.[72,73]

Extractions with a 0.1 mol L$^{-1}$ EDTA solution and various ascorbic acid concentrations (0.001, 0.01, 0.1 and 1 mol L$^{-1}$) were performed. The pH varied between 3.9 and 4.2 and redox potentials were of the order of 0.2 V (pe + pH ranging between 6.8 and 8.9). Despite the fact that these solutions present higher potential values than preceding ones (EDTA–TiCl$_3$ and EDTA–TGA), similar results were obtained (Table 14).

The solubilisation of lead which was not extracted by the 0.1 mol L$^{-1}$ EDTA solution on soils A, B and C is correlated to the dissolution of iron oxides with increasing ascorbic acid concentrations.

In the experiments where soils B and C were extracted using a 0.1 mol L$^{-1}$ EDTA–1 mol L$^{-1}$ AA solution, iron dissolution was lower than that obtained with a 0.1 mol L$^{-1}$ EDTA–0.1 mol L$^{-1}$ AA solution whereas lead extraction

**Table 13** *Pb and Fe extraction with different EDTA–TGA solutions*

| | | $EDTA\ 0.1\ mol\ L^{-1}$ | $EDTA\ 0.1\ mol\ L^{-1}$ $TGA\ 0.01\ mol\ L^{-1}$ | $EDTA\ 0.1\ mol\ L^{-1}$ $TGA\ 0.1\ mol\ L^{-1}$ | $EDTA\ 0.1\ mol\ L^{-1}$ $TGA\ 1\ mol\ L^{-1}$ |
|---|---|---|---|---|---|
| Soil A | Pb (%) | 60 | 73 | 95 | 110 |
| | Fe (%) | 2 | 10 | 14 | 38 |
| Soil B | Pb (%) | 4.5 | 24 | 101 | |
| | Fe (%) | 0.03 | 2 | 10 | |
| Soil C | Pb (%) | 30 | 29 | 101 | 97 |
| | Fe (%) | 0.3 | 3 | 10 | 21 |
| Soil E | Pb (%) | 68 | 67 | 63 | 66 |
| | Fe (%) | 15 | 6 | 13 | 26 |
| Soil T | Pb (%) | 49 | 53 | 54 | 56 |
| | Fe (%) | 8 | 19 | 18 | 32 |

**Table 14**  *Pb and Fe extraction with different EDTA–AA solutions*

| | | EDTA 0.1 mol $L^{-1}$ | EDTA 0.1 mol $L^{-1}$ AA 0.001 mol $L^{-1}$ | EDTA 0.1 mol $L^{-1}$ AA 0.01 mol $L^{-1}$ | EDTA 0.1 mol $L^{-1}$ AA 0.1 mol $L^{-1}$ | EDTA 0.1 mol $L^{-1}$ AA 1 mol $L^{-1}$ |
|---|---|---|---|---|---|---|
| Soil A | Pb (%) | 60 | 55 | 56 | 87 | 103 |
| | Fe (%) | 2 | 5.5 | 8 | 16 | 20 |
| Soil B | Pb (%) | 4.5 | 6 | 14.4 | 83 | 100 |
| | Fe (%) | 0.03 | 0.3 | 2 | 18 | 12 |
| Soil C | Pb (%) | 30 | 22 | 20 | 87 | 100 |
| | Fe (%) | 0.3 | 1 | 3 | 17 | 9.2 |
| Soil E | Pb (%) | 68 | * | * | 64 | 61 |
| | Fe (%) | 15 | * | * | 18.2 | 23 |
| Soil T | Pb (%) | 49 | * | 30 | 44 | 40 |
| | Fe (%) | 8 | * | 18.5 | 23 | 26 |

*The final pH of these solutions was greater than 6. As the other solutions were at pH 4.5, these results were not used

increased. It would seem that iron dissolved from iron oxides formed another solid during the experiment. Such results were observed during the study of soils having the highest iron contents and in the presence of elevated ascorbic acid concentrations. The solubility limit of iron ascorbate was probably exceeded in these experiments, however we were not able to find information in the literature to confirm this hypothesis.

In the case of soil E, lead does not seem to be bound to an iron compound soluble in a reducing medium. This is also in agreement with the low iron content of this sample. In the same way, lead extraction from soil T is not favoured by a decrease in the redox potential of the extractant. Therefore the existence of an important fraction of total lead linked to the residual fraction (in the order of 50%), already suggested with the EDTA–TiCl$_3$ and EDTA–TGA extractions is confirmed.

### 4.3.3.4   Conclusions

The association of EDTA to a reducing reagent seems a satisfactory way of determining of the reducible fraction of lead with this kind of sample.

Some of the reagents studied (EDTA–TGA and EDTA–AA) were also applied to evaluate the mobilisation of copper (data not presented here). The results show that the EDTA–TGA extraction gives systematically lower extraction yields than those obtained with the EDTA–AA solution. This result may be attributed to a low solubility of copper thioglycolate or to the formation of insoluble Cu compounds with disulfides, degradation compounds of TGA appearing during the experiments. On the other hand, the comparison of extraction yields obtained with 0.1 mol L$^{-1}$ EDTA–0.1 mol L$^{-1}$ AA and oxalate solutions (step 'MTF3a' of Scheme 1) shows a good agreement between the values obtained. So this EDTA–AA solution has an efficiency similar to that of the step MTF3a of Scheme 1. It allows the solubilisation of species bound to amorphous iron oxides or not well crystallised iron oxides.

Therefore we chose to use a 0.1 mol L$^{-1}$ EDTA–0.1 mol L$^{-1}$ AA extracting solution to evaluate the reducible fraction in further studies.

### 4.3.4   Elaboration of Simple Fractionation Schemes for the Evaluation of the Risks of Pollutant Mobilisation from Industrial and Mining Site Soils

Two simple fractionation schemes, specific for cationic and anionic elements, were elaborated on the basis of the whole set of observations presented above in order to evaluate more reliably and more rapidly the mobile fraction, the residual fraction and the potential risks.

### 4.3.4.1   Simple Fractionation Scheme Specific for Elements Giving Cationic Species in IPS

This simplified fractionation scheme, designed to deal with IPS and called hereafter SFSC, consists of two single parallel extractions (see Appendix 4):

- 0.1 mol L$^{-1}$ EDTA solution at pH 4.5. The corresponding fraction is called SFSC1
- 0.1 mol L$^{-1}$ ascorbic acid–0.1 mol L$^{-1}$ EDTA solution at pH close to 4.5. The difference between the result of this test and the first one is called fraction SFSC2.

The first extraction (SFSC1), using EDTA in a slightly acidic solution, allows to define for each of the soils treated the amount of metal easily soluble. This fraction includes the 'exchangeable' and 'bound to carbonates' fractions defined by the Tessier scheme and also cations that were 'adsorbed on iron oxides and organic matter'.

The second extraction, associating both EDTA and ascorbic acid, induces the dissolution of iron oxides and extracts associated pollutants. The amount of cationic pollutants corresponding to the 'reducible' fraction SFSC2 is then calculated by difference.

Fraction SFSC3 is evaluated by difference between the total concentration and the sum (SFSC1 + SFSC2). This fraction includes the 'residual' and the strongly-bound part of the 'associated to organic matter' fractions of the Tessier scheme. The mobilisation in the environment of the part of the pollutant concentration described by SFSC3 is unlikely at least at short term.

## Experimental Application

### Case of Lead

In the case of lead the repartitions described by Figure 4 were obtained.

In the case of soils A, E and T, this scheme clearly shows the high mobility potential of lead in presence of a complexing agent or during acidification of the soil. For soils B and C, Pb release is probably due to the dissolution of iron oxides and requires reducing conditions.

Table 15 compares the residual lead fractions obtained with the different schemes.

The comparison between these different results underlines the difficulty in defining the residual fraction in IPS with accuracy. Indeed, a different value is obtained with each fractionation scheme. In four of the five studied samples, the lowest value is obtained with the simplified scheme SFSC, in the case of soil E

**Table 15** *Comparison of the residual fractions of lead determined by application of Tessier scheme, Scheme 1 and the simplified scheme SFSC*

| Scheme | Soil A | Soil B | Soil C | Soil E | Soil T |
|--------|--------|--------|--------|--------|--------|
| Tessier | 85% | 92% | 91.5% | 24% | 92% |
| 1 | 40% | 55% | 19% | 25% | 73% |
| SFSC3 | 13% | 17% | 12% | 32% | 51% |

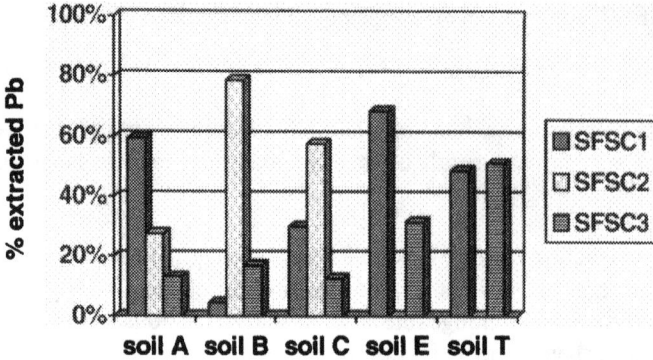

**Figure 4**   *Lead fractionation according to the simplified scheme SFSC*

the residual fraction is very slightly higher than those obtained with Scheme 1 or the Tessier scheme. This difference could be attributed to the non-extraction of lead strongly associated to the organic matter (7% according to the Tessier scheme).

## Case of Copper

The distribution obtained with the simplified scheme SFSC is reported in Figure 5.

In soil A copper is distributed between the three fractions. Comparing this repartition with the results of Scheme 1, it appears that the last fraction (SFSC3) is 10% superior to the residual fraction of Scheme 1. This result could be attributed to an incomplete extraction of copper associated to the organic matter during the first step (SFSC1). On the other hand the fractions soluble in reducing conditions are identical.

In the case of soil B, copper appears totally soluble in the reducing attack of SFSC whereas this fraction was evaluated only to 53% with Scheme 1. So this simplified scheme allows a better evaluation of the residual fraction.

The fraction SFSC3 of soil C amounts to 70% of total copper whereas the residual fraction of Scheme 1 is only of 29.5%. This difference may be attributed

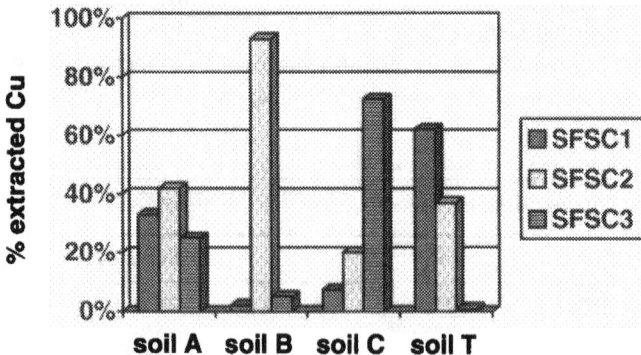

**Figure 5**   *Copper fractionation according to the simplified scheme SFSC*

to an incomplete dissolution of the well-crystallised iron oxides during the EDTA–AA extraction. Indeed, in Scheme 1, 25% of copper was extracted during the step MTF3c specific to crystallised iron oxides. So a better knowledge of the different iron oxides, such as that given by XRD, in the studied IPS samples seems necessary to evaluate well the residual fraction.

The application of the simplified scheme SFSC to soil T put in evidence the high copper mobility. The better part of this element is dissolved in an acid or complexing solution (60%) and the rest of Cu is soluble in a reducing solution. So Scheme 1 has led to the overestimation of the residual fraction (25%) and underestimation of the potential risk of mobilisation in a reducing environment.

## Conclusion

A two-step simplified scheme has been elaborated to determine in a short time the easily soluble fraction of cationic elements in IPS during an acidification and the fraction soluble in a reducing environment. It is based on the use of a 0.1 mol $L^{-1}$ EDTA solution, alone or associated with a reducing reagent, ascorbic acid. The latter allows the dissolution of iron oxyhydroxides and the release of the associated metallic pollutants.

The application of this scheme to the five IPS samples studied evidenced a correlation between iron dissolution and lead extraction in the case of soils B and C. For the other samples, 50% or more of the total lead is easily dissolved in acidic or complexing conditions and risk being redistributed in the environment. In the case of copper, the results are similar to those obtained with Scheme 1.

The proposed simple fractionation scheme SFSC seems to be suitable for an evaluation of the mobile fraction of the pollutants in the case of industrially-polluted soils and mining sites.

### 4.3.4.2   Simple Fractionation Scheme Specific to Arsenic (Anionic Elements)

Another simplified fractionation scheme, called hereafter SFSA, was developed to deal with the fractionation in IPS of elements giving anionic species, especially arsenic (see Appendix 4).

It consists of two successive extractions, first by a $KH_2PO_4/K_2HPO_4$ 0.1 mol $L^{-1}$ solution at pH 7.2, then by oxalate/oxalic acid 0.2 mol $L^{-1}$ solution at pH 2 completed by a determination of total As following alkaline fusion.

The first fraction (SFSA1) represents easily extractable As which is the most immediately dangerous for the environment. A potassium phosphate buffer solution (pH = 7.2) was already recommended by some authors to extract the exchangeable fraction, particularly for elements giving anionic species.[74] Indeed phosphate anions can compete for the exchange sites and reduce arsenic adsorption.[63,75,76]

The second step evaluates As extractable under moderately reducing and complexing conditions. This fraction (SFSA2) is susceptible of being mobilised in reducing environments where iron oxyhydroxides may be dissolved.

As in the simplified scheme for cations SFSC proposed above the last fraction, calculated by subtraction from total As of the sum (SFSA1 + SFSA2) does not strictly represent only 'residual' As as defined by the Tessier scheme. It includes both the part of As very strongly linked to organic matter, which could be released in very oxidising conditions, and that strongly bound in crystalline forms.

This simplified scheme SFSA was evaluated by application to the five IPS samples studied (Figure 6).

The aim of the first extraction is to evaluate the anionic exchangeable fraction SFSA1. Unlike what occurs during the corresponding $MgCl_2$ extraction used in the Tessier scheme or Scheme 1, pH conditions remain constant during the whole extraction procedure (pH = 7.2). According to the results shown in Figure 6, exchangeable As (SFSA1) in soil E is very high (35%), this confirms its high mobility in this soil. Although the extracted percentage is much lower for soils A and T (3.5% and 6% respectively), it must be noted that the As concentrations released are far from negligible (497 and 1751 mg kg$^{-1}$ respectively). So the risks of remobilisation by anion exchange on these two heavily polluted soils have also to be seriously taken into consideration.

By comparing the results obtained with the sequential extraction schemes (steps TF1 or MTF1), it appears that exchangeable fractions measured with the phosphate buffer extraction are higher than those obtained with $MgCl_2$. Indeed, $Mg^{2+}$ can not be exchanged with the anionic species and the univalent chloride ion has a low anion exchange power. Another weakness common to the Tessier scheme and Scheme 1 is that as the pH is not buffered during $MgCl_2$ extraction procedures; low final pH values are sometimes observed, particularly with IPS. This certainly contributes to attack the carbonated fraction and favours As extraction. So in some cases, the exchangeable fraction (TF1 or MTF1) obtained with the Tessier scheme or Scheme 1 could, on the contrary, be overestimated.

Therefore phosphate extraction seems to be more appropriate for the evaluation of easily extractable As and should be preferred for the characterisation of As mobility in industrial soils.

The majority of the arsenic content of studied IPS is leached in the second

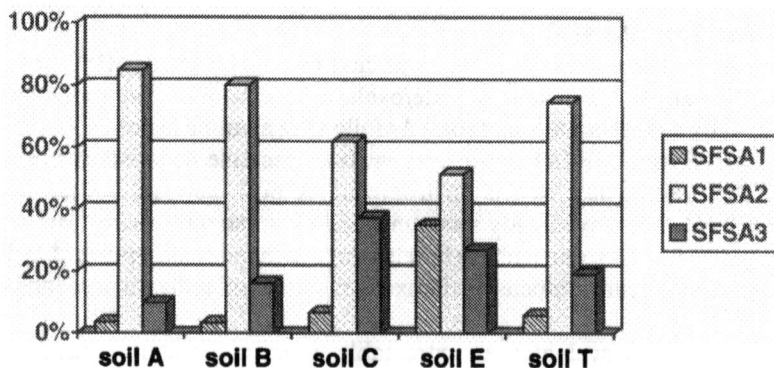

**Figure 6**    *Fractionation of As in IPS samples by the simplified scheme SFSA*

fraction SFSA2 (moderately reducing and complexing conditions, pe + pH ranging between 5.8 and 8). For each of the five IPS studied the residual As fraction SFSA3 determined (in the sequence soil C > soil E > soil B = soil T > soil A) is the same as that found with the eight-step sequential extraction scheme (soil C > soil E > soil B > soil T > soil A), indicating that this fraction is very likely representative of As strongly bound to the soil and very difficult to solubilise.

### 4.3.4.3 Application to a Certified Reference Sediment and Speciation Study

This simplified scheme SFSA was also applied to a certified BCR sediment (CRM 320) and a speciation method (HPLC–HG–AFS) used to analyse the extracts. The results summarised in Table 16 show a quantitative recovery of total As during the three steps.

The pollutant is preferentially mobilised in the moderately reducing and complexing fraction SFSA2 (64%) whereas only 24% is released in the phosphate exchangeable fraction SFSA1, in good agreement with the previous conclusions. In order to complete information about As behaviour in this CRM a speciation study of As in the supernatant liquids was also carried out using HPLC–HG–AFS[77] (Figure 7).

Most of the arsenic in the phosphate extract (92% of the As extracted in this step) was found as As(v), As(III) being only a minor component. During the oxalate extraction As(v) was the only detected form, this a proof that this extraction procedure does not proceed by inducing a reduction of As(v) to As(III), at least at the time scale of the experiment. Another conclusion of this speciation analysis is that whereas some As(III) may be found in exchangeable forms only As(v) is sorbed on iron oxides.

So, the easily extractable fraction and the fraction that may be dissolved in reducing conditions can be assessed with a simplified and rapid procedure, without modifying As speciation. This opens the field for further investigations combining fractionation by selective extractions to speciation studies.

**Table 16**  *As fractionation and speciation in CRM 320 according to the simplified scheme SFSA*

|                 | As(III)                      | As(v)                        | Sum                          |
|-----------------|------------------------------|------------------------------|------------------------------|
| SFSA1           | 1.5 mg kg$^{-1}$<br>2%       | 16.9 mg kg$^{-1}$<br>22%     | 18.4 mg kg$^{-1}$<br>24%     |
| SFSA2           | < d.l.                       | 49.0 mg kg$^{-1}$<br>64%     | 49.0 mg kg$^{-1}$<br>64%     |
| SFSA3           | /                            | /                            | 14.0 mg kg$^{-1}$<br>18%     |
| Total           | /                            | /                            | 81.4 mg kg$^{-1}$            |
| Certified value | /                            | /                            | 76.7 ± 3.4                   |

**Oxalate extraction**

**Phosphate extraction**

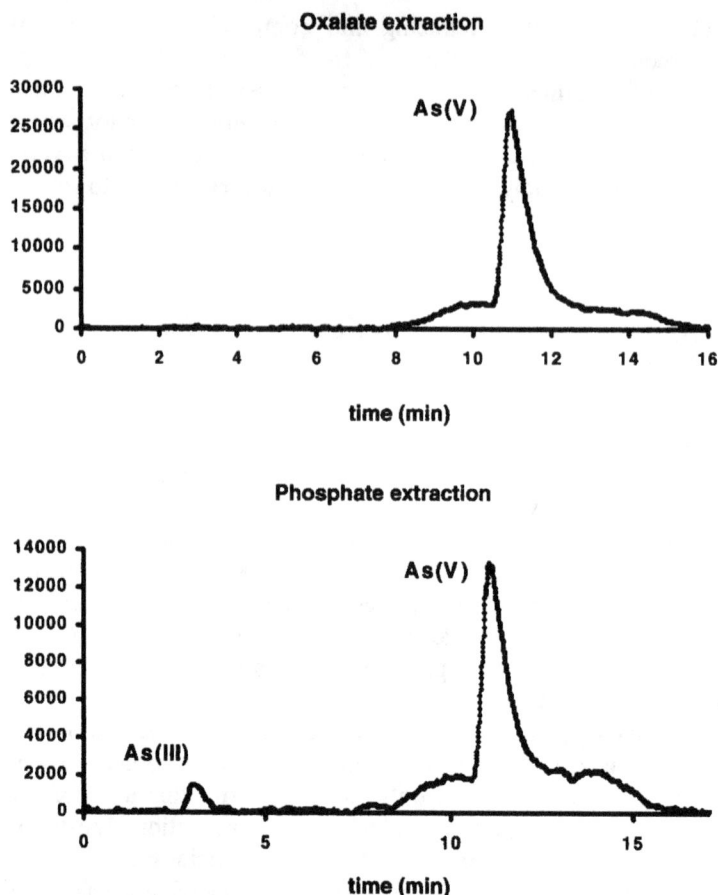

**Figure 7**   *HPLC–HG–AFS speciation of As in the phosphate and oxalate SFSA extracts of CRM 320*

### 4.3.4.4   *The Case of a Multiple Pollution*

The study of polluted soils containing elements giving both cationic and anionic species is not rare, especially when dealing with IPS. It is therefore necessary to possess methods allowing the characterisation of the mobility of the pollutants in all cases.

It is the reason why the simple specific scheme SFSC elaborated for fractionation studies of elements giving cationic species was also tested in the case of arsenic in order to generalise its application to a multiple pollution study.

The results presented in Figure 8 evidence the importance of reducing conditions in arsenic dissolution. They have to be compared to those obtained with the anion specific scheme SFSA, already presented in Figure 6.

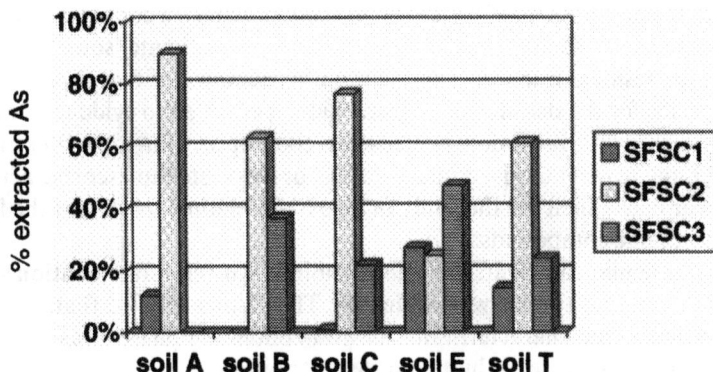

**Figure 8** *As fractionation according to the simplified scheme specific to cationic species, SFSC*

The two simplified schemes (specific to cationic species and specific to arsenic) give a similar evaluation of the residual fraction for soils A and T.

On the other hand this residual fraction is overestimated by the specific scheme of cationic species SFSC in the case of soils B (37% *vs* 20%) and E (47.5% *vs* 27%); this result was attributed for these two soils to an underestimation of the fraction soluble in reducing conditions during the EDTA–AA extraction. On the contrary the residual fraction appears somewhat underestimated by SFSC for soil C.

It is nevertheless evident from these results that a rough evaluation of As fractionation in the five IPS studied by the use of the simplified fractionation scheme specific of cationic species SFSC seems possible. It gives a first evaluation of the potential mobilisation of arsenic and so can be useful as a first step in the case of multiple pollution.

## 4.4 Conclusions

Sequential selective extraction techniques are commonly used to fractionate the solid phase forms of metals and metalloids in soils or sediments and so evaluate the potential mobilisation of these elements. Application to industrial or mining sites polluted soils (IPS) has not been the object of much attention.

In this study five samples originating from industrial and mining sites were sequentially extracted using the classical Tessier scheme and several others, adapted from this procedure (Scheme 1) or specifically designed (SFSC and SFSA). These samples are highly polluted by metals and arsenic and present very peculiar matrix characteristics, such as very high iron or carbonate contents, as compared to natural soils. However these complex matrices are quite common when dealing with IPS.

From this study it may be concluded that the fractionation scheme used must be adapted in function of the main characteristics of the studied sample (organic matter, carbonates, iron, calcium *etc.*) and of the pollutant content. The comparison between the results obtained with the different schemes used showed that, whatever the pollutant, the Tessier scheme does not allow a correct evaluation of the Fe-associated fraction in IPS and leads to overestimate, sometimes enormously, the residual fraction. The addition in Scheme 1 of two supplementary steps, specific for the dissolution of iron oxides, permitted to evidence a correlation between iron dissolution and arsenic and copper (for a part) extraction. However, the reagent used – oxalate – alone or associated to ascorbic acid does not appear convenient for the study of lead fractionation because of the formation of insoluble compounds.

In the particular case of arsenic, the possibility of using a fractionation scheme specific to cationic species was evaluated. The results showed that, as for the Tessier scheme, the characterisation of iron-containing phases was insufficient and could lead to a false evaluation of As distribution.

From these results, and those of a complementary study applying different reducing solutions, two simplified fractionation schemes were elaborated.

The first one, SFSC, is designed specifically for cationic species but can also be used in the case of a multiple pollution. It consists of two parallel experiments and presents the option to employ EDTA alone or associated to ascorbic acid to determine the easily mobilisable fraction and the fraction soluble in reducing conditions. The pH of the extracting solutions is not very acid (4.5) and, from this point of view, it allows a more realistic evaluation of pollutant mobilisation risks in the environment. On the other hand, the fraction incorporated in strongly crystallised iron oxides or strongly bound to the organic matter cannot be assessed individually, leading to attribute higher values to the residual fraction in some cases.

The second scheme, SFSA, was elaborated for the fractionation of anionic species and evaluates in two consecutive steps the phosphate-exchangeable fraction and the fraction soluble in reducing conditions. The use of phosphate as the ion exchange reagent improves much the evaluation of anionic forms, at least for arsenic.

From these results a first evaluation of the possible mobilisation of pollutants from IPS during pH or Eh changes can be realised. In all IPS samples studied, arsenic appears as an element easily mobilised when the environment becomes more reducing and presents then an elevated risk of transfer to the soil solution. The metallic cations are generally distributed essentially between the easily soluble and reducing fractions and, as for arsenic, acidification and moreover a lowering of redox potential could facilitate their mobilisation.

## 4.5   References

1. B.J. Alloway, *Proceedings of the 3rd International Conference on the Biogeochemistry of Trace Elements*, Paris, France, 1995.
2. A.C.M. Bourg in *Heavy Metals, Problems and Solutions*, F.a.M. Salomons (ed.), Springer, Berlin, 1995, 19.
3. D.M. Templeton, F. Ariese, R. Cornelis, L.-G. Danielsson, H. Muntau, H.P. Van Leeuwen and R. Lobinski, *Pure Appl. Chem.*, 2000, **72(8)**, 1453.
4. P. Cambier, *Analusis Magazine*, 1994, **22(2)**, 21.

5. D. Baize in *Teneurs Totales en Éléments Traces Métalliques dans les Sols*, Inra, Paris, France, 1997.
6. R. McLaren, *Proceedings of the 16th World Congress of Soil Science*, Montpellier, France, 1998.
7. A.B. Ribeiro and A.A. Nielsen, *Geoderma*, 1997, **76**, 253.
8. F. Persin, T. Toularastel, J. Sandeaux, R. Sandeaux and C. Gavach, *Proceedings of the 3rd International Conference on the Biogeochemistry of Trace Elements*, Paris, France, 1995.
9. P. Adamo, S. Dudka, M.J. Wilson and W.J. McHardy, *Environ. Pollut.*, 1996, **91(1)**, 11.
10. C. Gleyzes, PhD Thesis, Université de Pau et des Pays de l'Adour, France, 1999.
11. J. Thöming and W. Calmano, *Acta Hydrochim. Hydrobiol.*, 1998, **26**, 338.
12. J.E. Van Benschoten, M.R. Matsumoto and W.H. Young, *Journal of Environmental Engineering*, 1997, **March**, 217.
13. H.A. Elliot, J.H. Linn and G.A. Shields, *Hazardous Waste and Hazardous Materials*, 1989, **6(3)**, 223.
14. M.C. Steele and J. Pitchel, *Journal of Environmental Engineering*, 1998, **124(7)**, 639.
15. W.R. Berti and S.D. Cunningham, *Environ. Sci. Technol.*, 1997, **31**, 1359.
16. K. Hudson-Edwards, M. Macklin, C. Curtis and D. Vaughan, *Proceedings of the 3rd International Conference on the Biogeochemistry of Trace Elements*, Paris, France, 1995.
17. L.Q. Ma and G.N. Rao, *J. Environ. Qual.*, 1997, **26**, 259.
18. A. Chlopecka, J.R. Bacon, M.J. Wilson and J. Kay, *J. Environ. Qual.*, 1996, **25**, 69.
19. M. Pantsar-Kallio and P.K.G. Manninen, *Science Total Environ.*, 1997, **204**, 193.
20. K. Kalbitz and R. Wennrich, *Science Total Environ.*, 1998, **209**, 27.
21. A. Kabata-Pendias in *Heavy Metals, Problems and Solutions*, F.a.M. Salomons, (ed.), Springer, Berlin, 1995, 3.
22. C. Segues, PhD Thesis, Université de Pau et des Pays de l'Adour, France, 1998.
23. P.J. Kavanagh, M.E. Farago, I. Thornton and R.S. Braman, *Chemical Speciation and Bioavailability*, 1997, **9**, 77.
24. C.A. Jones, W.P. Inskeep and D.R. Neuman, *J. Environ. Qual.*, 1997, **26(2)**, 433.
25. C.P. Waller, R.P. Edwards and C. Wilkins, *Proceedings of the International Land Reclamation and Mine Drainage Conference and the Third International Conference on the Abatement of Acidic Drainage*, Pittsburgh, 1994, 244.
26. A. Lebourg, T. Sterckeman, H. Ciesielski and N. Proix, *Agronomie*, 1996, **16**, 201.
27. C. Juste, *Science du Sol*, 1988, **26**, 103.
28. A. Tessier, P.G.C. Campbell and M. Bisson, *Anal. Chem.*, 1979, **51**, 844.
29. L.M. Shuman, *Soil Sci.*, 1985, **140**, 11.
30. G. Rauret, J.F. López-Sánchez, A. Sahuquillo, R. Rubio, C.M. Davidson, A.M. Ure and Ph. Quevauviller, *J. Environ. Monit.*, 1999, **1**, 57.
31. W.F. Pickering, *Ore Geol. Rev.*, 1986, **1**, 83.
32. F.M.G. Tack and M.G. Verloo, *Intern. J. Environ. Anal. Chem.*, 1995, **59**, 225.
33. G. Rauret, *Talanta*, 1998, **46**, 449.
34. Ph. Quevauviller, G. Rauret, H. Muntau, A.M. Ure, R. Rubio, J.F. López-Sánchez, H.D. Fiedler and B. Griepink, *Fresenius J. Anal. Chem.*, 1994, **349**, 808.
35. G. Bombach, A. Pierra and W. Klemm, *Fresenius J. Anal. Chem.*, 1994, **350**, 49.
36. L.A. Legiec, L.P. Griffin, P.D. Walling, T.C. Breske, M.S. Angelo and R.S.A.S. Isaacson, *Environmental Progress*, 1997, **16(1)**, 29.
37. K.A. Gruebel, J.A. Davis and J.O. Leckie, *Soil Sci. Soc. Am. J.*, 1988, **52**, 390.
38. R.G. McLaren, R. Naidu, J. Smith and K.G. Tiller, *J. Environ. Qual.*, **27**, 1998, 348.

39. S.C. Chang and M.L. Jackson, *Soil Sci.*, 1957, **84**, 133.
40. E.A. Woolson, J.H. Axley and P.C. Kearney, *Soil Sci. Soc. Amer. Proc.*, 1973, **37**, 254.
41. V. Ruban, J.F. López-Sánchez, J.F. Pardo, G. Rauret, H. Muntau and Ph. Quevauviller, *J. Environ. Monit.*, 1999, **1**, 51.
42. B.M. Onken and D.C. Adriano, *Soil Sci. Soc. Am. J.*, 1997, **61**, 746.
43. Y. Zhang and J.N. Moore, *Environ. Sci. Technol.*, 1996, **30**, 2613.
44. D.A. Dzombach and F.M.M. Morel in *Surface Complexing Modeling. Hydrousferric Oxide*, Wiley-Interscience, New York, 1990.
45. P.H. Masscheleyn, R.D. Delaune and W.H. Patrick, *Environ. Sci. Technol.*, 1991, **25(8)**, 1414.
46. P.H. Masscheleyn, R.D. Delaune and W.H. Patrick, *J. Environ. Qual.*, 1991, **20**, 522.
47. H. Xu, B. Allard and A. Grimvall, *Water, Air, Soil Pollut.*, 1991, **57–58**, 269.
48. A. Spierenburg and C. Demanze, *Environnement et Technique*, Info-dechets courants, 1995, 79.
49. C. Delmas, Rapport de DEA, Université de Pau et des Pays de l'Adour, France, 1997.
50. P.P. Coetzee, *Water SA*, 1993, **21**, 291.
51. A. Barona and F. Romero, *Soil Technology*, 1996, **8**, 303.
52. R.B. Herbert Jr, *Water, Air, Soil Pollut.*, 1996, **96**, 39.
53. M. Raksasataya, A.G. Langdon and N.D. Kim, *Anal. Chim. Acta*, 1996, **332**, 1.
54. Z. Li and L. M. Shuman, *Environmental Pollution*, 1997, **95(2)**, 227.
55. M. Sadiq, *Water, Air, Soil Pollut.*, 1997, **93**, 117.
56. C. Gommy, PhD Thesis, Université de Technologie de Compiègne, France, 1997.
57. A.E. Martell and R.M. Smith in *Critical Stability Constants*, Plenum Press, New York, 1976.
58. R.C. Weast in *Handbook of Chemistry and Physics*, R.C. Weast, (ed.), CRC Press, Cleveland, Ohio, 1975–1976.
59. E. Jeanroy, PhD Thesis, Université de Nancy I, France, 1983.
60. A. Andrieu-Linares, *Physio-Géo.*, 1982, **4**, 71.
61. F. Trolard, Mémoire présenté pour l'obtention du grade d'Agrégé de l'Enseignement Supérieur, Université Catholique de Louvain, Belgium, 1997.
62. J.L. Gomez Ariza, I. Giraldez, D. Sanchez-Rodas and E. Morales, *Talanta*, 2000, **52**, 545.
63. H. Yan-Chu in *Arsenic in the Environment, Part I: Cycling and Characterisation*, O.J. Nriagu, (ed.), John Wiley & Sons, Inc., New York, 1994, 17.
64. N. Belzile and A. Tessier, *Geochim. Cosmochim. Acta*, 1990, **54**, 103.
65. A.L. Foster, G.E. Brown Jr, T.N. Tingle and G.A. Parks, *American Mineralogist*, 1998, **83**, 553.
66. F. Juillot, Ph. Ildefonse, G. Morin, G. Calas, A.M. de Kersabiec and M. Benedetti, *Applied Geochemistry*, 1999, **14**, 1031.
67. R. Williams, *Journal of Agricultural Science*, 1937, **27**, 259.
68. O.K. Borggaard, *Clay Minerals*, 1982, **17**, 365.
69. J. Schröeter and J. Thöming, Report, Technische Universität Hamburg-Harburg, Arbeitsbereich Umweltschutztechnik, Germany, 1999.
70. G. Heron, T.H. Christensen and J.C. Tjell, *Environ. Sci. Technol.*, 1994, **28**, 153.
71. L.M. Shuman, *Soil Sci. Soc. Am. J.*, 1982, **46**, 1099.
72. D. Suter, S. Banward and W. Stumm, *Langmuir*, 1991, **7**, 809.
73. W. Stumm in *Chemistry of the Solid–Water Interface in Natural Systems*, John Wiley & Sons, Inc., 1992.
74. J.M. Asikainen and N.P. Nikolaidis, *Ground Water Monitoring and Remediation*, 1994, **14**, 185.

75. N.T. Livesey and P.M. Huang, *Soil Sci.*, 1981, **131**, 88.
76. E.A. Crecelius, N.S. Bloom, C.E. Cowan and E.A. Jenne in *Ecological Studies Program, Energy Analysis and Environment Division EA-4641*, Vol. 2, 1986.
77. Y. Bohari, A. Astruc, M. Astruc and J. Cloud, *Journal of Analytical Atomic Spectrometry*, 2001, **16**, 774.

# Appendix 2; Sequential Extraction Procedures

All reagents should be of analytical grade or better.

## Tessier Scheme

*Exchangeable (TF1)*: 1 g of soil was agitated at room temperature for 1 h with 8 ml of 1 mol $L^{-1}$ $MgCl_2$ at pH = 7.

*Carbonates (TF2)*: the solid residue from TF1 was agitated with 8 ml of 1 mol $L^{-1}$ sodium acetate/acetic acid buffer at pH 4.5 for 15 h at room temperature.

*Reducible (TF3)*: the solid residue from TF2 was extracted with 20 ml of 0.04 mol $L^{-1}$ hydroxylamine hydrochloride in 25% (v/v) acetic acid at $96 \pm 5°C$ in a waterbath for 5 h 30.

*Organic matter (TF4)*: The residue from TF3 was extracted with 3 ml of 0.02 mol $L^{-1}$ nitric acid and 5 ml of 30% (v/v) hydrogen peroxide. The mixture was heated to $85 \pm 5°C$ in a waterbath for 2 h. A second aliquot of 3 ml of 30% $H_2O_2$ was then added and the mixture was heated at the same temperature for 3 h. After cooling, 5 ml of 3.2 mol $L^{-1}$ ammonium acetate in 20% (v/v) nitric acid were added. Then, the sample was diluted to 20 ml and agitated continuously for 30 m.

*Residual (TF5)*: The final solid residue was digested with a mixture of hydrofluoric and perchloric acids in Teflon beakers. 0.2 g of sample was first digested with a solution of concentrated $HClO_4$ (2 ml) and HF (10 ml) and evaporated to near dryness; subsequently a second addition of $HClO_4$ (1 ml) and HF (10 ml) was made and again the mixture was evaporated to near dryness. Finally $HClO_4$ (1 ml) alone was added and the sample was evaporated until the appearance of white fumes. The residue was dissolved in concentrated nitric acid and diluted to 25 ml.

After each extraction step, the suspensions were centrifuged for 30 m at 4000 rpm. The supernatant liquid was decanted and placed in an acid-washed tube. The solid residue was rinsed with 8 ml of deionised water. The rinse water was separated from the solids in a similar way. The supernatant liquid obtained at each step and the rinse water were analysed for arsenic and metals.

## Scheme 1

The steps MTF1, MTF2, MTF3, MTF4 and MTF5 are identical to the steps TF1, TF2, TF3, TF4 and TF5.

Two steps were added in order to separate the different iron oxides:

*Amorphous iron oxides (MTF3a):* the solid residue from MTF3 was agitated with 50 ml of 0.2 mol $L^{-1}$ oxalate/0.2 mol $L^{-1}$ oxalic acid for 4 h in the dark.

*Crystalline iron oxides (MTF3c):* the solid residue from MTF3a was extracted with 50 ml of 0.2 mol $L^{-1}$ oxalate/ 0.2 mol $L^{-1}$ oxalic acid/0.1 mol $L^{-1}$ ascorbic acid, in a boiling waterbath for 30 m.

## Scheme 2

*Water soluble (WS):* 2.5 g of soil was agitated at room temperature for 30 m with 50 ml water.

*Al-associated (Al):* the solid residue from step WS was agitated at room temperature for 1 h with 50 ml of 0.5 mol $L^{-1}$ ammonium fluoride adjusted to pH = 8.2. The solid residue was then rinsed with 25 ml of saturated sodium chloride.

*Fe-associated (Fe):* the solid residue from step Al was agitated at room temperature for 18 h with 50 ml of 0.1 mol $L^{-1}$ sodium hydroxide.

*Ca-associated (Ca):* the solid residue from step Fe was agitated at room temperature for 1 h with 50 ml of 0.5 mol $L^{-1}$ ammonium fluoride adjusted to pH = 8.2. The solid residue was then rinsed with 25 ml of saturated sodium chloride.

*Occluded Al-associated (AlO):* the solid residue from step Ca was agitated at room temperature for 18 h with 50 ml of 0.5 mol $L^{-1}$ ammonium fluoride adjusted to pH = 8.2. The solid residue was then rinsed with 25 ml of saturated sodium chloride.

*Residual fraction (R):* this was obtained by an alkaline fusion; 0.5 g of soil were mixed with 4.5 g of hydroxide sodium in a nickel crucible and heated for 15 minutes. After cooling, the solid residue was dissolved in water and diluted to 100 ml.

After each extraction step, the suspensions were centrifuged for 30 m at 4000 rpm. The supernatant liquid was decanted and placed in an acid-washed tube.

# Appendix 3; Single Extractions

## EDTA Extraction

1 g of soil was agitated at room temperature for 2 h with 100 ml of 0.1 mol $L^{-1}$ $Na_2EDTA$ solution. After extraction, the suspensions were centrifuged for 30 m at 4000 rpm. The supernatant liquid was decanted and placed in an acid-washed tube.

## EDTA–TiCl₃ Extraction

*Solution (a)*: 2 g of soil was agitated at room temperature for 1 h 30 with 50 ml of 0.1 mol $L^{-1}$ $Na_2EDTA$ solution.
*Solution (b)*: 2 g of soil was agitated at room temperature for 1 h 30 with 50 ml of 0.1 mol $L^{-1}$ $Na_2EDTA$/0.008 mol $L^{-1}$ $TiCl_3$ solution.
*Solution (c)*: 2 g of soil was agitated at room temperature for 1 h 30 with 50 ml of 0.1 mol $L^{-1}$ $Na_2EDTA$/0.0292 mol $L^{-1}$ $TiCl_3$ solution.

After each extraction (a, b and c) the suspensions were centrifuged for 15 m at 4000 rpm. The supernatant liquid was decanted and placed in an acid-washed tube. Each extraction was followed by a rinse step using the same solution (20 ml) for 10 min. The rinse liquid was separated from the solids in a similar way and mixed with the preceeding extracting solution.
*Solution (d)*: This extraction was carried out in a beaker (Pyrex) on a magnetic stirrer. 2 g of soil was agitated at room temperature for 1 h 30 with 50 ml of 0.1 mol $L^{-1}$ $Na_2EDTA$/0.008 mol $L^{-1}$ $TiCl_3$ solution. During this extraction, potential and pH were continuously controlled. Several small volumes of $TiCl_3$ and $NH_3$ solutions were added to keep reducing conditions and constant pH during the whole extraction process.

## EDTA–TGA Extraction

2 g of soil was agitated at room temperature for 24 h with 50 ml of 0.1 mol $L^{-1}$ $Na_2EDTA$/0.01, 0.1 and 1 mol $L^{-1}$ TGA solutions. After the extractions, the suspension was centrifuged for 30 min at 4000 rpm. The supernatant liquid was decanted and placed in an acid-washed tube.

In the case of the 0.1 mol $L^{-1}$ $Na_2EDTA$/0.1 mol $L^{-1}$ TGA and 0.1 mol $L^{-1}$ $Na_2EDTA$/1 mol $L^{-1}$ TGA solutions, TGA was partially neutralised with soda (soda concentration being half that of the TGA concentration).

## EDTA–AA Extraction

2 g of soil was agitated at room temperature for 24 h with 50 ml of 0.1 mol $L^{-1}$ $Na_2EDTA$/0.01, 0.1 and 1 mol $L^{-1}$ ascorbic acid solutions at pH close to 4.5.

In order to neutralise the first acidity of ascorbic acid, NaOH was added during the preparation of the solutions. After the extractions, the suspension was centrifuged for 30 m at 4000 rpm. The supernatant liquid was decanted and placed in an acid-washed tube.

# Appendix 4: Simplified Schemes

### Simple Fractionation Scheme Specific for Arsenic (Anionic Elements) called SFSA

*First step (SFSA1)*: 2 g of sample was agitated at room temperature for 16 h with 40 ml of a 0.1 mol $L^{-1}$ $KH_2PO_4$/0.1 mol $L^{-1}$ $K_2HPO_4$ solution (pH = 7.2).

*Second step (SFSA2)*: the residue of the first step was agitated at room temperature with 100 ml of 0.2 mol $L^{-1}$ oxalate/0.2 mol $L^{-1}$ oxalic acid for 4 h in the dark.

*Third step (SFSA3)*: the final solid residue was evaluated using an alkaline fusion; 2 g of soil was mixed with 4.5 g of hydroxide sodium in a nickel crucible and heated for 15 minutes. After cooling, the solid residue was dissolved in water and diluted to 100 ml.

After each extraction step, the suspensions were centrifuged for 30 m at 4000 rpm. The supernatant liquid was decanted and placed in an acid-washed tube. The solid residue was rinsed with 16 ml of deionised water. The rinse water was separated from the solids in a similar way. The supernatant liquid obtained at each step and the rinse water were analysed by the same method.

### Simple Fractionation Scheme Specific for Elements Giving Cationic Species called SFSC

This consists of two single direct extractions.

*Extraction 1*: 1 g of soil was agitated at room temperature with 100 ml of 0.1 mol $L^{-1}$ $Na_2EDTA$ solution at pH 4.5 for 2 h (pH = 4.5). After the extraction, the suspension was centrifuged for 30 m at 4000 rpm.

*Extraction 2*: 2 g of soil was agitated at room temperature with 50 ml of 0.1 mol $L^{-1}$ ascorbic acid/0.1 mol $L^{-1}$ $Na_2EDTA$ solution at pH close to 4.5 for 24 h. In order to neutralise the first acidity of ascorbic acid, 0.05 mol $L^{-1}$ NaOH was added during the preparation of the solution. After the extraction, the suspension was centrifuged for 30 m at 4000 rpm.

CHAPTER 5

# Sequential Extraction Procedures for Phosphorus Forms in Lake Sediment

V. RUBAN,[1] J.F. LÓPEZ-SÁNCHEZ, P. PARDO,[2]
G. RAURET, H. MUNTAU AND Ph. QUEVAUVILLER

[1]Laboratoire Central des Ponts et Chaussées, Division Eau, Bouguenais,
France
[2]University of Barcelona, Department of Analytical Chemistry, Barcelona,
Spain

## 5.1   Introduction

Eutrophication (*i.e.* the proliferation of algae due to an excess of nutrients) has become one of the major water pollution problems all over Europe, with strong economic consequences (*e.g.* the preparation of drinking water becomes more difficult because of filter clogging by algae). Phosphorus (P) is regarded as a key factor responsible for the eutrophication of fresh water.[1-5] Its concentration in lakes and rivers results both from external inputs and internal loading from the sediment, which can contribute phosphate to the overlying water at levels comparable to the external source;[6] its release depends on the form of phosphate in the sediment. Most of sediment phosphorus is in the particulate form, dissolved P being only a few percent of total P. Not all the forms of phosphorus are bioavailable and, therefore, likely to increase eutrophication. Although standardised methods to quantify total phosphorus in waste and fresh water exist, such a method is not yet available for the determination of different forms of phosphate; these forms can be determined using sequential extraction procedures which are operationally-defined methods related to specific reagents and procedures, *i.e.* results are interpreted to be related to a specific phase of sediment (although, *sensus stricto*, it is solely related to a chemical procedure). In the absence of standardised method(s), it is not possible to compare data from one laboratory to another nor to extrapolate results from one lake to another. At this stage, consequently, it becomes urgent to identify and standardise a harmonised method which would enable us to achieve comparability of data among monitoring laboratories and would hence be an essential tool for water managers.

In order to improve this situation, the European Commission through the Standards, Measurements and Testing programme has launched a collaborative project which aimed to (1) design a harmonised sequential extraction scheme, (2) test the selected scheme in interlaboratory studies involving expert European laboratories, and (3) certify the extractable phosphorus content of a sediment reference material. The project was started in 1996 by a literature search which identified four methods, among the most widely recognised, for the determination of phosphorus compounds in lake sediment; these were taken as a basis for comparison and for working out a harmonised procedure.

## 5.2    Selection of Sequential Extraction Procedures

Sequential extraction schemes were first developed for soils and then extended to sediments.[7] Operationally-defined schemes are available, most of them addressing inorganic forms of phosphate. The main inorganic forms of P are (i) the fraction adsorbed by exchange sites which are referred to as loosely bound, labile or exchangeable P.[8–10] This fraction is easily releasable and becomes available for algal growth; (ii) the fraction associated with Al, Fe and Mn oxides and hydroxides. Phosphorus and iron are often bound in sediments, P is then adsorbed onto iron complexes by ligand exchanges,[11] the amount of FeOOH is one of the factors controlling P release from the sediment; (iii) the fraction in Ca-bound compounds generally referred to as apatite-P[12,13] or Ca-bound.[14,15] P adsorption onto calcium carbonate is one of the mechanisms for the formation of calcium-P in sediments. However, contrarily to the formation of iron-P, Ca-bound P can also be formed by precipitation. Organic P is a complex fraction the exact nature of which is not precisely known yet. De Groot and Golterman[16] have shown that organic-P is partly composed of phytates. Among the four sequential extraction schemes selected for evaluation in the frame of this project, three rely on 'strong' reagents, namely the procedures developed by Williams,[12] Hieltjes–Lijklema[8] and Ruttenberg;[9] whereas chelators are used in the method developed by Golterman;[14] these procedures are summarised in Table 1.

The Williams and Hieltjes–Lijklema methods use NaOH to solubilise Fe and HCl to dissolve Ca; in the case of the Hieltjes–Lijklema method, the NaOH extraction is preceded by $NH_4Cl$ extraction. The shortcomings of these methods have already been outlined.[17,18] The Williams method leads to resorption by carbonates in calcareous sediments whereas dissolution of small amounts of Fe–P and Al–P by $NH_4Cl$ is possible in the Hieltjes–Lijklema scheme.

NaOH/HCl is replaced by EDTA in the Golterman method; this is supported by the suspicion that strong acids and alkaline solutions are too aggressive and should be avoided in extraction schemes. In this method, chelating agents such EDTA would extract Fe-bound and Ca-bound phosphorus without disturbing clay-bound or organic phosphorus. The Golterman method is not without pitfalls and the determination of phosphate can be perturbed by interference generated by EDTA.

**Table 1** Reagents used and corresponding P fractions in the four sequential extraction procedures tested in this study

| | Step 1 | Step 2 | Step 3 | Step 4 | Step 5 |
|---|---|---|---|---|---|
| Williams[12] | NaOH 1 mol L⁻¹ iron-bound P bioavailable | HCl 1 mol L⁻¹ + 3.5 mol L⁻¹ Ca-bound P non available | HCl 3.5 mol L⁻¹ + calcination total P | HCl 1 mol L⁻¹ + calcination organic P partly available | |
| Hieltjes–Lijklema[8] | NaH₄Cl 1 mol L⁻¹ labile P bioavailable | NaOH 0.1 mol L⁻¹ + 2 mol L⁻¹ iron-bound P bioavailable | HCl 0.5 mol L⁻¹ Ca-P non available | | |
| Golterman[14] | H₂O labile P bioavailable | Ca-EDTA dithionite 0.05 mol L⁻¹ iron-P bioavailable | Na₂-EDTA 0.1 mol L⁻¹ Ca-P non available | H₂SO₄ 0.25 mol L⁻¹ acid soluble organic P bioavailable | NaOH 2 mol L⁻¹ reductant organic P non available |
| Ruttenberg[9] | MgCl₂ 1 mol L⁻¹ loosely sorbed P bioavailable | Na₃ citrate 0.3 mol L⁻¹ and NaHCO₃ 1 mol L⁻¹ Fe-bound P bioavailable | Na-acetate 1 mol L⁻¹ authigenic apatite, Ca-bound P, biogenic apatite non available | HCl 1 mol L⁻¹ detrital apatite non available | HCl 1 mol L⁻¹ + calcination organic P partly available |

The Ruttenberg method, initially developed for marine sediments, is a five-step procedure which allows the separation of the following sedimentary P pools: loosely sorbed or exchangeable; ferric Fe-bound P; authigenic carbonate fluoroapatite + biogenic apatite + $CaCO_3$ associated P; detrital apatite; and organic P. It is the first method designed to chemically separate authigenic apatite from detrital apatite. Furthermore, the problem of analytical artefacts resulting from redistribution of P onto residual solid surfaces during extraction has been resolved. The advantages and disadvantages of each method are listed in Table 2.

The opportunity to test the BCR scheme developed for sequential extraction of trace elements in sediments has also been discussed. This method is a three-step procedure in which acetic acid, hydroxylammonium chloride and hydrogen peroxide are successively used.[19,20] The first step of this method appeared to be inadequate for P extraction and, consequently, the method was not selected for the preliminary trial.

# 5.3   Interlaboratory Study

## 5.3.1   Samples and Analytical Techniques Used in the Interlaboratory Study

Almost all laboratories applied colorimetry as the final detection method for the phosphorus extracted by the different procedures (based on the Murphy and Riley spectrophotometric method).[21] ICPAES has also been used but was not suitable since this technique enables the determination of total phosphate. Spectrophotometry determines soluble reactive phosphorus whereas ion chromatography determines orthophosphate only.

All the reagents used were analytical grade reagents; the glassware and plastic

**Table 2**   *Advantages and disadvantages of the four selected sequential extraction procedures*

| Methods | Advantages | Disadvantages |
|---------|------------|---------------|
| Williams | Simple, practical | Partial resorption of P extracted by NaOH on $CaCO_3$ |
| Hieltjes–Lijklema | Simple, practical | Dissolution of small amounts of Fe–P and Al–P by NaOH; hydrolysis of organic P; no relation with bioavailability |
| Golterman | Extracts specific compounds; permits extraction of organic P fractions; provides information on bioavailable fractions | Not practical; NTA and EDTA interfere with P determination; complicated solution preparation; in some sediments, extraction must be repeated |
| Ruttenberg | Distinction between different apatite forms; no redistribution of P onto residual solid surfaces during extraction | Very long, not practical; the butanol extraction is very difficult to achieve |

ware were soaked in HCl 0.3% and rinsed with deionised water. A quality control procedure was applied throughout the different steps of sample preparation and analysis.

In order to identify and eliminate the causes of discrepancies between laboratories, the first intercomparison started with the analysis of an extract of a real material. The trial then focused on two sediments to be analysed by the four above described methods. For each method, a detailed protocol was set out and had to be strictly followed by the participants. A second trial was carried out, in which three sediments with different characteristics, *i.e.* organic, siliceous and calcareous, were analysed using one method selected by the laboratories among the four schemes previously tested.

Lake sediments S23, S24, S2, S10A and S22 were used for the study; they were collected and prepared (air-died at room temperature, sieved at 90 $\mu$m, homogenised and bottled) by the JRC at Ispra. The sediments were tested for their homogeneity and total P (determined by XRF) was found to be homogeneous in all the samples. The composition of the samples are given in Table 3.

## 5.3.2 Discussion of the Results of the First Trial

### 5.3.2.1 Williams Method

Table 4a shows the mean of laboratory means with the SD and coefficient of variation (CV) for the different steps of each method, for sediments S23 and S24. These results are commented on in Table 5. This method does not pose any particular problem and gives good analytical results; it is simple to use compared to the Ruttenberg and Golterman methods. In both cases the average value for total P was very close to that determined by ICP. However, there are some pitfalls, *e.g.* organic P could be slightly underestimated due to losses during calcination.

### 5.3.2.2 Golterman Method

The results are presented in Table 4b and commented on in Table 5. Several labs complained about the imprecision and duration of this method. Furthermore, one can note that for all the laboratories, the sum of the different fractions was always less than total P (TP) determined by ICP.

**Table 3** *Composition of the sediments used in the trials*

|      | Si % | Al % | Ca % | Fe % | P % | Org. C % |
|------|------|------|------|------|------|------|
| S23  | 18.5 | 7.3  | 11.6 | 4.7  | 0.058 | 1.6 |
| S24  | 11.9 | 3.8  | 19   | 1.11 | 0.096 | 4.1 |
| S2   | 12.5 | 3.7  | 3.1  | 4.9  | 0.523 | 20.3 |
| S10A | 19.1 | 4.6  | 1.9  | 3.9  | 0.343 | 10.4 |
| S22  | 11.4 | 4.7  | 21.4 | 1.4  | 0.086 | 1.2 |

**Table 4a**    *Results of the first interlaboratory trial – Williams protocol*

| Extracted forms of P | Number of sets | Mean ± SD (mg kg$^{-1}$ P) | CV (%) raw data |
|---|---|---|---|
| Sediment S23 | | | |
| HCl-P | 6 | 338 ± 16 | 4.7 |
| NaOH-P | 6 | 92 ± 2 | 20.7 |
| Organic P | 6 | 154 ± 11 | 7.2 |
| Inorganic P | 6 | 442 ± 14 | 3.2 |
| Conc. HCl-P | 6 | 636 ± 19 | 3.0 |
| Sediment S24 | | | |
| HCl-P | 6 | 352 ± 14 | 4.0 |
| NaOH-P | 6 | 165 ± 24 | 14.5 |
| Organic P | 6 | 180 ± 12 | 6.7 |
| Inorganic P | 6 | 508 ± 23 | 4.5 |
| Conc. HCl-P | 6 | 771 ± 44 | 5.7 |

**Table 4b**    *Results of the first interlaboratory trial – Golterman protocol*

| Extracted forms of P | Number of sets | Mean ± SD (mg kg$^{-1}$ P) | CV (%) raw data |
|---|---|---|---|
| Sediment S23 | | | |
| Step 1 | 6 | 3 ± 2 | 66.7 |
| Step 2 | 6 | 203 ± 33 | 16.3 |
| Step 3 | 6 | 19 ± 4 | 21.1 |
| Step 4 | 6 | 47 ± 26 | 55.3 |
| Step 5 | 6 | 82 ± 40 | 48.8 |
| Sediment S24 | | | |
| Step 1 | 6 | 10 ± 11 | 110 |
| Step 2 | 6 | 190 ± 55 | 29.0 |
| Step 3 | 6 | 239 ± 11 | 19.3 |
| Step 4 | 6 | 63 ± 6 | 9.5 |
| Step 5 | 6 | 166 ± 65 | 39.2 |

### 5.3.2.3    Hieltjes–Lijklema Method

The results of the interlaboratory trial on sediment S23 and S24 are reported in Table 4c. Though this method seemed simple, involving the same reagents as the Williams method, the results of the trial were not very good (see comments in Table 5).

### 5.3.2.4    Ruttenberg Method

Similar to the Golterman method, this method is time consuming and the many steps are a source of cumulative error in spite of the detailed protocol. The results are presented in Table 4d and commented in Table 5.

**Table 4c**    *Results of the first interlaboratory trial – Hieltjes–Lijklema protocol*

| Extracted forms of P | Number of sets | Mean ± SD (mg kg P$^{-1}$) | CV (%) raw data |
|---|---|---|---|
| Sediment S23 | | | |
| Step 1 | 6 | 32 ± 7 | 21.9 |
| Step 2 | 6 | 99 ± 18 | 18.2 |
| Step 3 | 6 | 147 ± 142 | 96.5 |
| Sediment S24 | | | |
| Step 1 | 6 | 99 ± 17 | 17.2 |
| Step 2 | 6 | 116 ± 20 | 17.2 |
| Step 3 | 6 | 326 ± 26 | 8.0 |

**Table 4d**    *Results of the first interlaboratory trial – Ruttenberg protocol*

| Extracted forms of P | Number of sets | Mean ± SD (mg kg$^{-1}$ P) | CV (%) raw data |
|---|---|---|---|
| Sediment S23 | | | |
| Step 1 | 6 | 44 ± 17 | 38.6 |
| Step 2 | 6 | 156 ± 294 | 186 |
| Step 3 | 6 | 117 ± 60 | 51.3 |
| Step 4 | 6 | 143 ± 40 | 27.9 |
| Step 5 | 6 | 139 ± 19 | 13.7 |
| Sediment S24 | | | |
| Step 1 | 6 | 96 ± 15 | 15.6 |
| Step 2 | 6 | 38 ± 97 | 156 |
| Step 3 | 6 | 233 ± 86 | 63.1 |
| Step 4 | 6 | 95 ± 25 | 26.3 |
| Step 5 | 6 | 170 ± 39 | 22.9 |

## 5.3.3    Reasons for Adopting the Williams Extraction Scheme

The benefit of undertaking the trial was twofold (i) the trial showed the large difference in the results obtained for the same sediment using different extraction schemes and, therefore, confirm the necessity of developing a harmonised procedure, (ii) it also showed that the modified Williams protocol seemed to be the most promising method for achieving comparability of results.

Following this first trial, and by common consent both from the EC Commission and the laboratories, it was decided to reject the Ruttenberg and Golterman protocols. Though these methods might present a few advantages, *e.g.* extraction of different organic P fractions, information on bioavailable fractions (Golterman), distinction between different apatite forms (Ruttenberg), they are not reproducible and they are difficult to carry out. Consequently, they were not considered to be suitable for adoption as a harmonised procedure, which should be both reliable and precise. A discussion started over the Williams and Hieltjes–Lijlema methods. Though the same reagents are used in both protocols, their concentrations differ; they are higher in Williams, which seems better.

**Table 5**    *Comments on the results of the first sediment trial*

| Williams | NaOH-P | Fairly high spreading of the results both for S23 and S14 (CV ≈ 20%) |
|---|---|---|
| | HCl-P | Reproducibility is good for both sediments |
| | Organic P | Reproducibility is good for both sediments |
| | Inorg. P | Reproducibility is good for both sediments |
| | Conc. HCl-P | Reproducibility is good for both sediments. In all cases, the average value for total P is close to that determined by ICP |
| Hieltjes–Lijklema | Step 1 | Reproducibility is not very good (CV ≈ 18%). |
| | Step 2 | For both sediments, reproducibility is not very good. |
| | Step 3 | Large spreading of the results for S23. Results were much better for S24 after the neutralisation step was suppressed |
| Golterman | Step 1 | Poor results partly due to the very low amount of P extracted |
| | Step 2 | For both sediments, reproducibility is not good. The results depend on the number of extractions performed |
| | Step 3 | For both sediments, reproducibility is not good. The results depend on the number of extractions performed |
| | Step 4 | The results improved for S24 due to recommendations (definition of the number of extractions, improvement of shaking) |
| | Step 5 | Reproducibility is not good (CV ≈ 40%) |
| Ruttenberg | Step 1 | Reproducibility is not good for both sediments |
| | Step 2 | Critical step due to the butanol extraction (CV > 150%) |
| | Step 3 | Reproducibility is low for both sediments |
| | Step 4 | Reproducibility is low for both sediments |
| | Step 5 | This step yields the best results of the method (CV ≈ 17%) |

Williams yields organic P which is not the case for Hieltjes–Lijlema (organic P can be calculated as the difference between total P and inorganic P). Hieltjes–Lijlema yields soluble P, this phase is not determined in the Williams scheme. Besides, Hieltjes–Lijlema is a sequential extraction scheme, whereas in the Williams protocol the steps are independent, not sequential. It was also pointed out that the Williams method gave better results in the trials and the participants agreed to take this protocol as a working basis for the harmonised procedure.[22]

## 5.3.4   Discussion of the Results of the Second Trial

The modified Williams scheme was then tested on three sediments with different characteristics *i.e.* organic, siliceous and calcareous (Table 3). The results of this trial are reported in Table 6.

**Table 6** *Results of the second interlaboratory study on the modified Williams protocol*

| Extracted forms of P | Number of sets | Number of accepted sets | Mean ± SD (raw data) (mg kg⁻¹ P) | CV (%) raw data | CV (%) accepted data |
|---|---|---|---|---|---|
| Sediment S2 | | | | | |
| NaOH-P | 15 | 13 | 2055 ± 560 | 27 | 15 |
| HCl-P | 15 | 14 | 615 ± 25 | 41 | 16 |
| Organic P | 10 | 10 | 886 ± 60 | 7 | 7 |
| Inorganic P | 10 | 9 | 3052 ± 470 | 15 | 7 |
| Conc. HCl-P | 10 | 9 | 4248 ± 160 | 4 | 3 |
| TP (digestion) | 13 | 12 | 5219 ± 4270 | 82 | 11 |
| Sediment S10 | | | | | |
| NaOH-P | 15 | 12 | 1044 ± 190 | 18 | 8 |
| HCl-P | 15 | 14 | 388 ± 150 | 39 | 9 |
| Organic P | 10 | 10 | 594 ± 30 | 5 | 5 |
| Inorganic P | 10 | 10 | 1461 ± 100 | 7 | 7 |
| Conc. HCl-P | 10 | 9 | 2304 ± 130 | 6 | 4 |
| TP (digestion) | 13 | 12 | 2901 ± 2340 | 81 | 4 |
| Sediment S22 | | | | | |
| NaOH-P | 15 | 14 | 88 ± 20 | 23 | 16 |
| HCl-P | 15 | 14 | 328 ± 140 | 43 | 7 |
| Organic P | 10 | 9 | 106 ± 10 | 9 | 5 |
| Inorganic P | 10 | 9 | 406 ± 60 | 15 | 4 |
| Conc. HCl-P | 10 | 8 | 538 ± 80 | 15 | 11 |
| TP (digestion) | 13 | 12 | 724 ± 590 | 81 | 5 |

The statistical tests of Cochran (95%) and Grubbs were applied to the raw data in order to eliminate laboratories with a high variance and outliers, respectively. The results were generally good (CV < 10% in most cases) and only a few labs had problems, which were solved. A good quality control procedure was highly recommended. No difference due to the nature of the sediment could be seen, consequently, the method should be suitable for all sediments.

The digestion step, which is not part of the Williams protocol, was carried out to test its advantages/disadvantages as regards with total P measured by the Williams method. This method is more difficult to handle than the Williams method because of the use of HF, which is not a routine analysis in many labs. Moreover, the results (accepted data) were very similar and it was decided not to use the digestion method and to stick to the Williams protocol.

One laboratory used ICP as the determination method, the results were systematically higher for NAIP, consequently the spectrophotometric determination was recommended. It has to be noted that the SMT protocol might not be the best with regard to the evaluation of the bioavailability of the different P fractions but is it certainly the most suited one to allow laboratories to achieve reproducible and comparable results. Moreover, the SMT method is simple to handle and could provide a useful tool for water managers on a routine basis, in the field of lake restoration. Especially, the method could help in calculating the releasable P stock in a lake sediment, hence information on the lake recovery delay.

Following the discussion of the results, slight modifications were suggested in order to improve the protocol and make it as clear as possible, even for laboratories who are not familiar with sequential extraction schemes. The good results allowed the modified protocol to be adopted for the final step of the project *i.e.* the certification campaign.

## 5.4 Certification Campaign

The modified Williams protocol, named SMT-protocol (see Appendix 5) has been reviewed by all participants and accepted as the common procedure for the certification of a lake sediment reference material which was conducted within the first half of 1999.

### 5.4.1 Preparation and Characterisation of the Reference Material

The reference material used as candidate CRM is siliceous, with low calcium and organic matter contents (Table 7). The total P content is around 0.126% (X-ray fluorescence determination). The sediment has been sampled in a shallow bay of an oligotrophic lake receiving large quantities of P. This material has been chosen because it contains low Ca and organic matter contents which makes analysis easier with respect to extractable forms of P.

### 5.4.2 Homogeneity and Stability Studies

#### 5.4.2.1 Homogeneity Study

The homogeneity of the material has been tested for both within- and between-bottle homogeneity and was found to be suitable for certification.[23,24] For the within-bottle homogeneity study, the forms of phosphorus were determined in the candidate CRM by analysing ten subsamples taken from one bottle, whereas one subsample in each of twenty different bottles selected during the bottling procedure was analysed for the between-bottle homogeneity test. The CVs and the total uncertainty $U_{CV}$ for the extractable P contents, between ($CV_B$) and within ($CV_W$) bottles, are given in Table 8.

**Table 7** *Mean composition of the candidate reference material (determined by XRF)*

|        | Si % | Al % | Ca % | Fe % | P % | Org. C % |
|--------|------|------|------|------|-------|------|
| Mean   | 23.7 | 7.89 | 5.62 | 4.84 | 0.126 | 2.66 |
| SD     | 0.30 | 0.11 | 0.06 | 0.07 | 0.002 | 0.04 |
| CV (%) | 1.27 | 1.35 | 1.11 | 1.47 | 1.83  | 1.56 |

**Table 8** *Within- and between-bottle variances resulting from the homogeneity study. Between-bottle: 20 replicate determinations; within-bottle: 10 replicate determinations*

| | Between-bottle CV (%) | Within-bottle CV (%) | | | |
|---|---|---|---|---|---|
| | | 1 | 2 | 3 | 4 |
| NaOH-P | 2.5 | 2.2 | 4.2 | 1.7 | 1.8 |
| HCl-P | 2.7 | 2.7 | 3.8 | 3.3 | 1.9 |
| Inorganic P | 1.6 | 1.6 | 1.4 | 1.5 | 1.0 |
| Organic P | 1.9 | 2.9 | 2.6 | 3.1 | 2.0 |
| Conc. HCl-P | 1.5 | 2.1 | 1.6 | 1.7 | 2.0 |

No difference was detected between the within- and between-homogeneity variances using an F-test (Table 9). Therefore, the material was considered to be homogeneous.

## 5.4.2.2 Stability Study

The stability of the extractable phosphate contents was tested to determine the suitability of the sediment as a reference material. Sets of bottles were kept at + 4, + 20 and + 40°C during a period of 12 months starting in December 1998; the extractable P contents were determined (in five replicates) after 1, 3, 6 and 12 months. The detection technique used is the same as in the homogeneity study.

Any change in the content of an analyte with time indicates an instability provided that a good long-term analytical reproducibility is obtained. Instability would be detected by comparing the contents of different analytes in samples stored at different temperatures with those stored at a low temperature at the various occasions of analysis. The experience has shown that storing sediment samples at − 20°C could result in changes of extractability, which was suspected for trace elements in a sediment reference material.[24] Consequently, it was decided to store the samples at + 4°C and use this temperature as reference for the samples stored at + 20 and + 40°C, respectively. Table 10 gives the ratios

**Table 9** *F-test used for the homogeneity study*

| | F experimental (F-test for comparison of variances) | | | |
|---|---|---|---|---|
| | B-W 1 | B-W 2 | B-W 3 | B-W 4 |
| NaOH-P | 1.303 | 2.769 | 2.113 | 1.887 |
| HCl-P | 1.046 | 2.095 | 1.567 | 1.919 |
| Inorganic P | 1.025 | 1.361 | 1.107 | 2.813 |
| Organic P | 2.204 | 1.782 | 2.570 | 1.042 |
| Conc HCl-P | 1.991 | 1.277 | 1.308 | 1.877 |

Critical values for a two-tailed test ($P = 0.05$): $F(19,9) = 3.68$; $F(9,19) = 2.88$
B = between-bottle variances
W = within-bottle variances

**Table 10**     *Stability tests of BCR 684*

| Month | NaOH-P | HCl-P | Inorganic P | Organic P | Conc HCl-P |
|---|---|---|---|---|---|
| | *Ratios value obtained/reference value, $R_T \pm U_T$, (samples stored at 20°C)* | | | | |
| 1 | $1.03 \pm 0.02$ | $0.98 \pm 0.03$ | $0.98 \pm 0.01$ | $0.97 \pm 0.03$ | $1.00 \pm 0.02$ |
| 3 | $1.01 \pm 0.01$ | $1.00 \pm 0.02$ | $1.01 \pm 0.01$ | $1.00 \pm 0.03$ | $1.01 \pm 0.01$ |
| 6 | $0.98 \pm 0.02$ | $0.94 \pm 0.02$ | $0.98 \pm 0.01$ | $0.98 \pm 0.01$ | $1.01 \pm 0.01$ |
| 12 | $0.98 \pm 0.01$ | $0.99 \pm 0.01$ | $1.02 \pm 0.01$ | $0.98 \pm 0.02$ | $1.03 \pm 0.02$ |
| | *Ratios value obtained/reference value, $R_T \pm U_T$, (samples stored at 40°C)* | | | | |
| Month | NaOH-P | HCl-P | Inorganic P | Organic P | Conc HCl-P |
| 1 | $1.01 \pm 0.02$ | $1.01 \pm 0.03$ | $0.97 \pm 0.02$ | $0.98 \pm 0.03$ | $1.00 \pm 0.02$ |
| 3 | $1.00 \pm 0.03$ | $1.02 \pm 0.02$ | $1.01 \pm 0.01$ | $1.00 \pm 0.02$ | $1.00 \pm 0.01$ |
| 6 | $1.00 \pm 0.01$ | $0.97 \pm 0.04$ | $0.99 \pm 0.01$ | $0.96 \pm 0.01$ | $1.01 \pm 0.01$ |
| 12 | $0.94 \pm 0.02$ | $1.01 \pm 0.01$ | $1.01 \pm 0.01$ | $0.93 \pm 0.02$ | $1.00 \pm 0.02$ |

($R_T$) of the mean values ($X_T$) of five measurements made at $+20$ and $+40°C$, respectively, and the mean value ($X_{+4°C}$) from five determinations made at the same occasion of analysis on samples stored at a temperature of $+4°C$: $R_T = X_T / X_{+4°C}$.

The uncertainty $U_T$ has been obtained from the coefficient of variation (CV) of five measurements obtained at each temperature: $U_T = (CV_T^2 + CV_{+4°C}^2)^{1/2} \times R_T/100$. In the case of ideal stability, the ratios $R_T$ should be 1. In practice, however, there are some random variations due to the error in the measurement. As indicated in Table 10, in most of the cases, the value 1 falls between $R_T - U_T$ and $R_T + U_T$. The uncertainty in the method can account for the deviations observed (in particular, a slight drift observed for the results obtained at $+4°C$ may result in variations of $R_T$; this is the case *e.g.* for organic P after 12 months). The material is considered to be stable and is recommended to be stored at $+4°C$.

## 5.4.3   Technical Discussion

### 5.4.3.1   General Comments

Each laboratory that took part in the certification exercise was requested to make a minimum of five independent replicate determinations of each extractable P form on at least two different bottles of the reference material on different days, following strictly the sequential extraction protocol described in Appendix 5. Details on the calibrants and calibration procedures are described in detail elsewhere.[23] A critical point stressed was the necessary calibration using extracting solutions (or external calibration with cross-check of calibrants in the extracting solutions).

All laboratories used colorimetry as final determination. A wavelength of 880 nm was originally specified in the protocol. However, two laboratories used a wavelength of 700 nm and did not observe any difference with the other results based on a 800 nm wavelength; consequently, it was decided to leave the choice of the wavelength open in the protocol.

The temperature of extraction was also discussed. A temperature of $21 \pm 1\,°C$ was required in the original protocol. It was pointed out that several laboratories had lower or higher temperatures without their results being affected. Therefore, the revised protocol was made a bit more flexible and now stipulates a temperature of $21 \pm 3\,°C$.

### 5.4.3.2   Results of the Trial

*NaOH-P*: The mean of laboratory means was $561\,mg\,kg^{-1}$ with a CV of 9.8% which was found to be acceptable. Suspicion was, however, thrown on apparently low (Lab 4) and high (Labs 13 and 16) results which were withdrawn from the overall set of data. Certification was based on 12 laboratory sets (Table 11). The resulting CV (6%) showed a great improvement in comparison to the second trial (CV of 20%).

*HCl-P*: Lab 4 had high results as a consequence of the low value found for NaOH-P and was consequently withdrawn. The certification was based on 14 laboratory data sets (Table 11). The CV (9%) also indicated a drastic improvement.

*Inorganic P*: The results were in excellent agreement with a CV of 3.5% which is actually in the range of the between-bottle variance. Certification was proposed on the basis of all laboratory sets (15 laboratories), Table 11.

*Organic P*: Lab 16 mentioned a measurement error which was due to an insufficient reaction time; indeed, the time of reaction to allow the colour to develop was shortened by 30 minutes, which emphasised the necessity to strictly stick to the Murphy and Riley method, *i.e.* allowing a reaction time of 2 h. The certification was based on 14 laboratory data sets, Table 11.

*Concentrated HCl-P*: An excellent agreement was obtained and certification was based on the 15 laboratory data sets (Table 11).

**Table 11**   *Certified values of extractable forms of P in the BCR 684*

| Extracted forms of P | Number of sets | Number of accepted sets | Mean ± 95% CI (raw data) (mg kg⁻¹ P) | CV (%) raw data (based on SD) | CV (%) Accepted data (based on 95% CI) |
|---|---|---|---|---|---|
| Sediment S2 | | | | | |
| NaOH-P | 15 | 12 | $550 \pm 21$ | 9.8 | 3.8 |
| HCl-P | 15 | 14 | $536 \pm 28$ | 9.4 | 5.2 |
| Inorganic P | 15 | 15 | $1113 \pm 24$ | 3.5 | 2.2 |
| Organic P | 15 | 14 | $209 \pm 9$ | 20.6 | 4.3 |
| Conc. HCl-P | 15 | 15 | $1373 \pm 35$ | 4.5 | 2.5 |

## 5.5   Conclusions

The stepwise approach followed to select the most suited extraction scheme for the determination of extractable phosphorus forms in fresh water sediment and to validate it in the frame of successive interlaboratory studies has proven to be very efficient in terms of agreement between the participating laboratories. The final agreement is indeed excellent which demonstrates that the SMT-protocol (modified Williams scheme) has a great potential for being accepted as a standard method by laboratories working in this area. Besides this 'pre-normative' work, the certification of a reference material certified for its extractable forms of P shows that (dried) fresh water sediments remain stable over long period of time which is a supplementary argument in favour of the proposed extraction scheme. Finally, the availability of this CRM will offer a great support to laboratories which will use this scheme, in terms of method validation and quality control.

## 5.6   References

1  R. Vollenweider, *Tech. Report OECD*, 1968, **27**, 59.
2.  OECD, *Synth. Report*, 1982, 164.
3.  H.L. Golterman and N.T. de Oude, *The Handbook of Environmental Chemistry*, 1991, 5A, 79.
4.  G. Barroin, *La Recherche*, 1991, **238**, 1412.
5.  C.P. Boers and D.T. Van der Molen, *Wat. Poll. Contr.*, 1993, **2**, 19.
6.  G. Rossi and G. Premazzi, *Wat. Res.*, 1991, **5**, 567.
7.  S.C. Chang and M.L. Jackson, *Soil Sci.*, 1957, **84**, 133.
8.  A.H.M. Hieltjes and L. Lijklema, *J. Environ. Qual.*, 1980, 3, 405.
9.  K.C. Ruttenberg, *Limnol. Oceanogr.*, 1992, 7, 1460.
10.  R. Psenner and R. Puckso, *Archiv. Hydrolbiol.*, 1988, **30**, 43.
11.  W. Stumm and J.J. Morgan, *Aquatic Chemistry: An Introduction Emphasising Chemical Equilibria in Natural Waters*, Wiley Interscience Publication, 2nd ed., 1981, 780.
12.  J.D.H. Williams, J.-M. Jaquet and R.L. Thomas, *J. Fish. Res. Bd. Can.*, 1976, **33**, 413.
13.  J.D.H. Williams, J.K Syers and T.W. Walters, *Soil Sci. Soc. Amer. Proc.*, 1967, **31**, 736.
14.  H.L. Golterman, *Hydrobiol.*, 1996, **5**, 1.
15.  H.L. Golterman, *Hydrobiol.*, 1982, **92**, 683.
16.  C.J. de Groot and H. Golterman, *Hydrobiol.*, 1993, **1**, 117.
17.  K. Pettersson, B. Boström and O.S. Jacobsen, *Hydrobiol.*, 1988, **170**, 91.
18.  A. Barbanti, M.C. Bergamini, F. Frascari, S. Miserocchi and G. Rosso, *J. Environ. Qual.*, 1994, **23**, 1093.
19.  A. Ure, Ph. Quevauviller, H. Muntau and B. Griepink, *Intern. J. Environ. Anal. Chem.*, 1993, **51**, 135.
20.  Ph. Quevauviller, G. Rauret, J.-F. López-Sánchez, R. Rubio, A. Ure and H. Muntau, *Sci. Total Environ.*, 1998, **205**, 223.
21.  J. Murphy and J.P. Riley, *Anal. Chim. Acta*, 1962, **27**, 31.
22.  V. Ruban, J.F. López-Sánchez, P. Pardo, G. Rauret, H. Muntau and Ph. Quevauviller, *J. Environ. Monitor.*, 1999, **1**, 51.
23.  V. Ruban, J.F. López-Sánchez, G. Rauret, H. Muntau and Ph. Quevauviller, Report EUR 19776 EN, European Commission, 43.
24.  P. Pardo, J.F. López-Sánchez, G. Rauret, V. Ruban, H. Muntau and Ph. Quevauviller, *Analyst*, 1999, **124**, 407.

# Appendix 5; SMT-protocol for Phosphorus Sequential Extraction in Fresh Water Sediments

## Reagents

HCl 1 mol L$^{-1}$
HCl 3.5 mol L$^{-1}$
NaOH 1 mol L$^{-1}$
NaCl 1 mol L$^{-1}$

## Apparatus

Centrifuge, balance
Polyethylene/polypropylene centrifuge tubes
Porcelain crucibles
Test tubes
4 mL and 20 mL pipettes
Shaker (*e.g.* magnetic stirrer, shaker table, end-over-end shaker)

## Remarks

1. The Murphy and Riley colorimetric method is used for phosphate determination.
2. All the reagents used must be of high purity grade (99.5% minimum).
3. All glassware and plasticware is to be cleaned in phosphate-free detergent, soaked for 24 h in 1.2 mol L$^{-1}$ HCl and air-dried before use.
4. Dry a separate 1 g sample of the sediment in an oven at 105 °C for 2 h and weigh. From this a correction to dry mass is obtained and applied to all analytical values reported (quantity per g dry sediment).
5. A blank sample should be carried out through the complete procedure for each extraction step.

6. The calibrant solutions should be made up with the appropriate extracting solutions.
7. The extraction should be carried out at a temperature of $21 \pm 3\,°C$.

# SMT Sequential Extraction Procedure for the Determination of Phosphorus Forms

## A   NaOH-extractable P and HCl-extractable P

I.   Weigh 200 mg of dry sediment in a centrifuge tube. It is important to keep the sediment/volume ratio constant. 200 mg of sediment is the minimum required

II.   Add with a pipette, 20 mL NaOH 1 mol $L^{-1}$

III.   Cover the tube and stir overnight (16 h). A good mixing is necessary, the sediment must be kept in suspension (use *e.g.* a magnetic stirrer, shaker table *etc.*)

IV.   Centrifuge at 2000 g for 15 minutes

### a   NaOH-P

1. Collect the supernatant
2. Set apart (with a pipette) 10 mL of the extract in a test tube
3. Add 4 mL HCl 3.5 mol $L^{-1}$
4. Stir energically for 20 seconds and let stand over night (16 h). Cover the tube
5. A brown precipitate appears and progressively settles. Centrifuge at 2000 g for 15 minutes
6. NaOH-P is measured in the supernatant by phosphate determination following the Murphy and Riley colorimetric method

### b   HCl-P

1. Wash the cake of the previous centrifugation (part A-IV) with 12 mL NaCl 1 mol $L^{-1}$. Stir for 5 minutes
2. Centrifuge at 2000 g for 15 minutes, discard the supernatant
3. Repeat parts b-1 and b-2 once
4. Add with a pipette 20 ml HCl 1 mol $L^{-1}$
5. Cover the tube and stir overnight (16 h)
6. Centrifuge at 2000 g for 15 minutes
7. HCl-P is measured in the supernatant by phosphate determination following the Murphy and Riley colorimetric method

## B   Concentrated HCl-extractable P

I.   Weigh 200 mg of dry sediment in a porcelain crucible
II.   Calcinate at 450°C for 3 h

III.   Pour the cool ash into a centrifuge tube
IV.   Add 20 mL HCl 3.5 mol L$^{-1}$ with a pipette. HCl can be added directly into the crucible to ease the transfer of the ash
V.    Cover the tube and stir overnight (16 h)
VI.   Centrifuge at 2000 g for 15 minutes
VII.  Collect the supernatant in a test tube for the analysis of concentrated HCl-P according to the Murphy and Riley colorimetric method

## C   IP (Inorganic Phosphorus) and OP (Organic Phosphorus)

### a   IP

1. Weigh 200 mg of dry sediment in a centrifuge tube
2. Add with a pipette 20 mL HCl 1 mol L$^{-1}$
3. Cover the tube and stir overnight (16 h)
4. Centrifuge at 2000 g for 15 minutes
5. Collect the extract in a test tube for IP analysis according to the Murphy and Riley colorimetric method

### b   OP

1. Add 12 mL demineralised water to wash the residue. Stir for 5 minutes
2. Centrifuge at 2000 g for 15 minutes, discard the supernatant
3. Repeat parts b-1 and b-2 once
4. Let the residue dry (in the tubes) in a ventilated drying cupboard at 80 °C. Put the tubes in an ultrasonic bath for 10 seconds and transfer to a porcelain crucible
5. Calcinate at 450 °C over 3 h
6. Pour the cool ash into the centrifuge tube
7. Add 20 mL HCl 1 mol L$^{-1}$ with a pipette. HCl can be added directly into the crucible to ease the transfer of the ash
8. Cover the tube and stir overnight (16 h)
9. Centrifuge at 2000 g for 15 minutes
10. Collect the extract in a test tube for OP determination following the Murphy and Riley colorimetric method

## Calculation

The concentration C, in mg g$^{-1}$ of P (dry weight) is:

$$C = \frac{S \times V}{1000\, m}$$

with: $S = P$ concentration in the extract (IP, OP, HCl-P) in mg $L^{-1}$ of P

    $V =$ volume of reagent used for extraction (20 mL)

    $m =$ mass of the test sample (200 mg dry weight)

For NaOH-P;

$$C = \frac{S \times 14 \times V}{1000 \times 10 \times m}$$

# Reference

J. Murphy and J.P. Riley, 'A modified single method for the determination of phosphate in natural waters', *Anal. Chim. Acta*, 1962, **27**, 31–36.

CHAPTER 6

# Leaching Procedure for the Availability of Polycyclic Aromatic Hydrocarbons (PAHs) in Contaminated Soil and Waste Materials

R.N.J. COMANS AND G.D. ROSKAM

Netherlands Energy Research Foundation (ECN), Petten, The Netherlands

## 6.1 Introduction

Analogous to inorganic contaminants, the environmental impact of organic pollutants in soil or waste materials is related to their 'availability' for transport and biouptake, rather than their total concentrations. Environmental risk analysis should, therefore, be based on (availability for) leaching of these contaminants, from the solid phase into solution. However, leaching tests for organic contaminants are not available at present. The objective of this chapter is to investigate the processes that control the leaching of polycyclic aromatic hydrocarbons (PAHs), as a common and typical category of hydrophobic organic contaminants, from contaminated soil and waste materials. The obtained insight in the leaching processes and controlling factors is used as the basis for the development of an 'availability' leaching test that is intended to indicate the maximum amount of the organic contaminants that can be leached from soil or waste materials. This chapter is based largely on work performed in the framework of two EU projects on the development of leaching tests for organic contaminants,[1,2] and on groundwater risk assessment at contaminated sites.[3]

Hydrophobic organic contaminants have a low solubility and bind strongly to both (natural) particular organic matter in the soil/waste matrix and dissolved organic matter in solution/leachate.[4] Figure 1 shows a simplified picture of the partitioning of PAHs between the particulate and water phase. As this figure

123

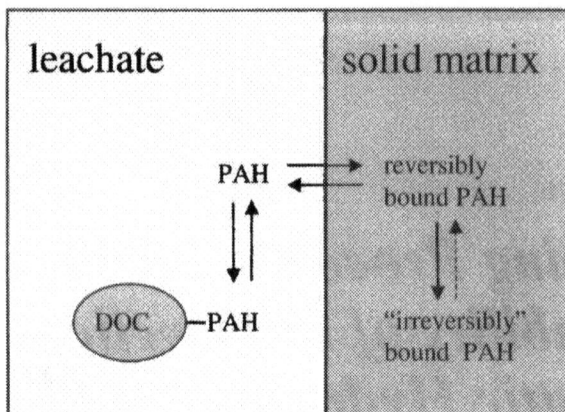

**Figure 1** *Partitioning processes controlling the leaching of organic contaminants from a soil or waste matrix*

clearly illustrates, the mobility and leaching of these organic contaminants is controlled by their fraction in the solid phase that is reversibly bound (*i.e.* 'available' for leaching) and by their solubility in both 'free' or truly dissolved and complexed or DOC-associated forms.

For example, the distribution of organic contaminants was studied among the solid phase and a solution phase that contained DOC (dissolved organic carbon).[5] The study concluded that the partitioning of strongly hydrophobic organic substances ($K_{ow} > 10^5$) is highly influenced by both the nature and the concentration of the dissolved organic macromolecules. The binding of these contaminants to DOC can thus result in a strong increase in their water-solubility and, therefore, facilitate their leaching from soil or waste materials.

Because of the importance of the role of dissolved organic carbon in enhancing the leaching of PAHs, the association of PAHs with DOC as well as the nature and behaviour of DOC are given particular emphasis in this study. Different methods are investigated to distinguish between truly dissolved and DOC-bound PAHs in leachates from contaminated materials; a modified 'solubility enhancement' method using [14]C-labelled pyrene and a modified Al-flocculation method to measure PAHs in leachates before (truly dissolved + DOC-bound) and after (only truly dissolved) the removal of DOC. Based on the dominating role of DOC in facilitating the leaching of PAHs, a procedure based on this process is evaluated as a potential 'availability' leaching test.

## 6.2 Materials and Methods

### 6.2.1 Materials

Three materials contaminated with PAHs have been used for this study, a typical PAH-contaminated soil and two waste materials:[2] a tar-containing asphalt granulate and the mechanically-biologically separated organic-rich fraction of municipal solid waste (referred to as OF-MSW). The soil sample originates from a former gasworks site. It was sampled from a location at the site that was considered not to be directly contaminated with coal tar, but with an intermediate contamination level of approximately 500 mg kg$^{-1}$ total-PAH. The OF-MSW material was chosen because it contains relatively high levels of PAHs and leaches very high concentrations of dissolved organic carbon (DOC). The tar-containing asphalt granulate originates from old road covers and contains very high PAH levels, but leaches relatively low concentrations of DOC. These three samples are believed to cover a wide spectrum of soil/waste materials with high PAH content and low (asphalt granulate), intermediate (gasworks soil) and high (OF-MSW) DOC concentrations in the leachates. Moreover, the properties of DOC in the three materials are likely to be different from one another. Therefore, this selection of samples is believed to facilitate a wide variety of observed leaching processes and applicability of the 'availability' test in the field of soil/sediment and waste materials.

## 6.2.2   Extraction and Analysis of PAHs

Two methods of PAH extraction and analysis have been used in this study: (1) C$_{18}$ solid phase extraction in combination with analysis by HPLC and (2) liquid/liquid extraction with dichloromethane in combination with (optimised) analysis by HPLC. We believe that Method 2 provides more reliable results over a wide range of leachate compositions.[1]

### 6.2.2.1   Method 1: C$_{18}$ Solid Phase Extraction + HPLC

C$_{18}$ solid phase extraction tubes (Supelco, SupelcleanTM LC-18 3 mL SPE tubes) were used for the extraction of PAHs from the leachates and were first pre-conditioned with methanol (Merck, Lichrosolv for liq. chrom.) and a 10% methanol/water solution. Prior to percolation through the extraction tubes, 10% methanol was added to the leachates to improve contact with the hydrophobic C$_{18}$ material. For the extraction of the PAHs from the column, 1 mL of acetonitrile (Merck, isocratic grade for liq. chrom.) and subsequently 3 × 0.5 mL of dichloromethane (Merck, GR for analysis) were used. The dichloromethane was evaporated under a gentle flow of nitrogen, resulting in about 1 mL of acetonitrile. The PAHs were analysed by HPLC with a C$_{18}$ PAH Column (Supelco, Supelcosil LC-PAH; length 5 cm, 4.6 mm internal diameter, particle size stationary phase 3 $\mu$m) and a mobile phase consisting of an acetonitrile/water mixture. Both a diode array and a fluorescence detector were used for the detection. This method was only used for the initial pH-static experiments (left hand graphs in Figure 3).

### 6.2.2.2   Method 2: Liquid/Liquid Extraction with DCM +
###              Optimised HPLC

PAHs were extracted by a three-step sequential liquid/liquid extraction with dichloromethane (water:DCM ratios were 75:10, 75:5 and 75:5, for 16, 3 and 3 h, respectively). The extract was dried by adding sodium sulfate (Merck, anhydrous granulated for organic trace analysis). The volume of DCM was reduced to 0.5 mL under a gentle flow of nitrogen and transferred to an HPLC-vial. Subsequently, 1 mL of acetonitrile was added and the volume was reduced to 1 mL by evaporating the DCM. If necessary (in the case of a turbid extract), a final clean-up of the DCM extract by percolation through aluminium oxide tubes (Supelco, SupelcleanTM LC-Alumina-N 6 mL SPE tubes) was added to the procedure. The PAHs were analysed by HPLC, using a $C_{18}$ PAH Column (Varian, Chromspher PAH; length 20 cm, 3 mm internal diameter, particle size stationary phase 5 $\mu$m), and a mobile phase consisting of an acetonitrile/water mixture. Concentrations of individual PAHs were measured by a fluorescence detector and confirmed by a diode array detector, except for acenaphthylene, which can only be detected by the diode array.

## 6.2.3   pH-Static Leaching

Because the leaching of DOC is generally strongly dependent on pH, batch pH-static leaching experiments were performed to investigate the effect of DOC on the leaching of PAHs from the soil and waste materials. The pH was monitored and adjusted to a set point in 250 mL glass reactors simultaneously at six pH values in the range of pH 4–13. Suspensions of the gasworks soil, OF-MSW and asphalt granulate were prepared in nanopure deionised water at a liquid/solid (L/S) ratio of 10 L kg$^{-1}$.

The first pH-static experiment with the gasworks soil was performed at a L/S ratio of 2 L kg$^{-1}$. The glass reactors were kept open to the atmosphere at $20 \pm 1\,°C$ and were stirred continuously during a 48 hour equilibration period, using a glass-coated magnetic stirring bar. The pH of the suspensions was continuously monitored by a computerised pH-static system and was automatically adjusted to the set point by addition of 1 mol L$^{-1}$ analytical grade HNO$_3$ or NaOH when the measured pH deviated by more than a pre-set value from the set point. In most cases it was necessary to adjust the highest pH-values (*i.e.* 12–13) manually by addition of 5 mol L$^{-1}$ NaOH.

After 48 h reaction time, the suspensions were pre-centrifuged at 2000 g for 0.5 h in 200 mL glass centrifuge tubes to facilitate the separation of leachate and soil/waste particles. Subsequently, the leachates were transferred to 40 mL Teflon-FEP centrifuge tubes and centrifuged at 27 000 g for 0.5 h, after which the clear supernatants were transferred to glass bottles. All glassware was cleaned by prior heat treatment at 550 °C.

The leachates were analysed for DOC by a Shimadzu TOC 5000A carbon analyser. PAH extraction and analysis were performed as described above. Size exclusion chromatography (SEC) was used to separate and characterise organic DOC molecules in the leachates on the basis of their size. The column that was used for the size-separation was a Waters Protein Pak 125 modified silica column, with a silica-based packing. A UV-absorbance detector (254 nm) and a Siever TOC-analyser were used for detection.

## 6.2.4  Soxhlet Extraction

10 g of solid sample was blended with 10 g of sodium sulfate (Merck, anhydrous granulated for organic trace analysis) and placed in an extraction thimble. For OF-MSW, a smaller sample size of 6.5 g was used. Due to the large amount of sodium sulfate that was needed to dry the sample, a larger sample would not fit in the extraction thimble. A glass wool plug was inserted on top of the solid material. 300 mL of dichloromethane (Merck, GR for analysis) was placed in a glass 500 mL round-bottom flask. The samples were extracted for 24 h at about 6 cycles per hour. After the extraction was completed, the extract was dried by percolation through a funnel filled with sodium sulfate. The volume was reduced to about 80 mL under a gentle nitrogen flow. Subsequently, 50 mL of acetonitrile (Merck, isocratic grade for liq. chrom.) was added, and the mixture was evaporated to 25 mL to ensure the removal of all dichloromethane. Acetonitrile was then added to a total volume of 200 mL, which was used for making the necessary dilutions for the HPLC-analysis.

## 6.2.5  DOC-Flocculation

Complexation–flocculation experiments were based on a recently developed method[6] with some modifications. The modified method has been applied to leachates obtained from extraction of the three soil/waste samples with $1 \text{ mol L}^{-1}$ NaOH, which resulted in leachates with a pH of between 13 and 13.5. Due to the high DOC concentrations in the OF-MSW extract, it was decided to add the large amounts of aluminium sulfate that were required to flocculate the DOC (*i.e.* 1, 10 and 100 g Al $\text{L}^{-1}$, except for asphalt granulate which was treated with 1, 10 and 25 g Al $\text{L}^{-1}$) as a solid salt rather than as a concentrated solution. Moreover, the pH of the solution was not set to pH 6 in advance (as in ref. 6), because the large amounts of aluminium sulfate already decreased pH sufficiently. In order to determine the effect of pH reduction, one batch of each sample was adjusted to pH 6 without addition of aluminium sulfate. In addition, one batch was adjusted to pH 1 in order to precipitate the humic acids.[7] After pH adjustment, the solutions were centrifuged at 27 000 g for 0.5 h and analysed by SEC and (after extraction of the PAHs) by HPLC as described above.

## 6.2.6  Development of the 'Availability' Leaching Test

The procedure of the developed 'availability' leaching test is described in detail elsewhere.[1] The availability for leaching is determined by extracting a sample of the ground (95% < 1 mm) soil/waste material with a solution of a commercial (Aldrich) humic acid (1000 mg C per L) at a high liquid/solid (L/S) ratio of 100 L $\text{kg}^{-1}$ and a pH of 12. This high pH-value is necessary to keep the DOC in solution by preventing its adsorption to the solid matrix of the soil/waste material. The concentration of DOC is chosen as an upper limit that can be reached, *e.g.* when alkaline materials react with organic soils.

In initial experiments, the effect of L/S ratio on the leaching of PAHs in the availability test was investigated by carrying out the test at L/S 10, 20, 50, 100, 500 and 1000 L kg$^{-1}$. The used amounts of dry soil/waste sample and leachate (1000 mg L$^{-1}$ Aldrich humic acid solution) were 20 g/200 mL, 10 g/200 mL, 4 g/200 mL, 5 g/500 mL, 2 g/1000 mL and 2.5 g/2500 mL, respectively. The effect of leaching time was studied by performing tests at L/S 100 and leaching times of 2 days, 4 days and 7 days.

## 6.3  Results

### 6.3.1  Total PAH Content of the Soil and Waste Materials

Figure 2 and Table 1 show the results of the Soxhlet extraction of the gasworks soil, the separated organic fraction from municipal solid waste (OF-MSW), and the asphalt granulate. The soil and waste samples contain substantial amounts of PAHs, with Σ16 US EPA priority PAH of 845 ± 31, 598 ± 26, and 3383 ± 53 mg kg$^{-1}$ for the gasworks soil, OF-MSW and asphalt granulate, respectively.

### 6.3.2  pH-Stat Leaching

Figure 3 shows concentrations of ΣPAH and DOC in pH-stat leachates of the gasworks soil, OF-MSW and asphalt granulate as a function of leachate pH.[2] Both the soil and the waste materials show a steep increase in DOC solubility at pH > 11, with levels of up to 60 000 mg kg$^{-1}$ in the OF-MSW leachates. The concentrations of ΣPAH in the leachates show a similar increase towards alkaline pH, which agrees well with the often reported role of DOC in the solubility enhancement of hydrophobic organic contaminants in natural aquatic systems.[4] It is noteworthy that the waste materials show similar features in the solubility/leaching behaviour of both DOC and PAHs as observed in natural (soil/sediment) systems.

The effect of DOC on the leaching of PAHs is more pronounced in the earlier experiments in which C$_{18}$ solid phase extraction was used to recover PAHs from the leachates. It is at present unclear whether this effect is due to the different extraction procedures (*i.e.* C$_{18}$ solid phase extraction *vs* dichloromethane liquid/liquid extraction) that have been used. Possibly, the recovery of the (more hydrophilic) DOC-associated PAHs is lower for the liquid/liquid extraction method than for the C$_{18}$ solid phase extraction. The noticeable physical retention of large DOC-molecules by the C$_{18}$ mini-columns that have been used may also have contributed to this effect.

Although no plateau is noticeable in the PAH leaching curve at alkaline pH (Figure 3), as is generally observed for inorganic contaminants (at low pH for metals and at high pH for anionic contaminants), we have taken the amount of PAHs that are leached at pH 12 as a first estimate of the fraction that is available for leaching. This pH value was chosen rather than pH 13 (*i.e.* the highest pH that was used in the experiments) because the latter pH often resulted in an incomplete recovery of PAHs and non-reproducible results. Moreover, measuring such

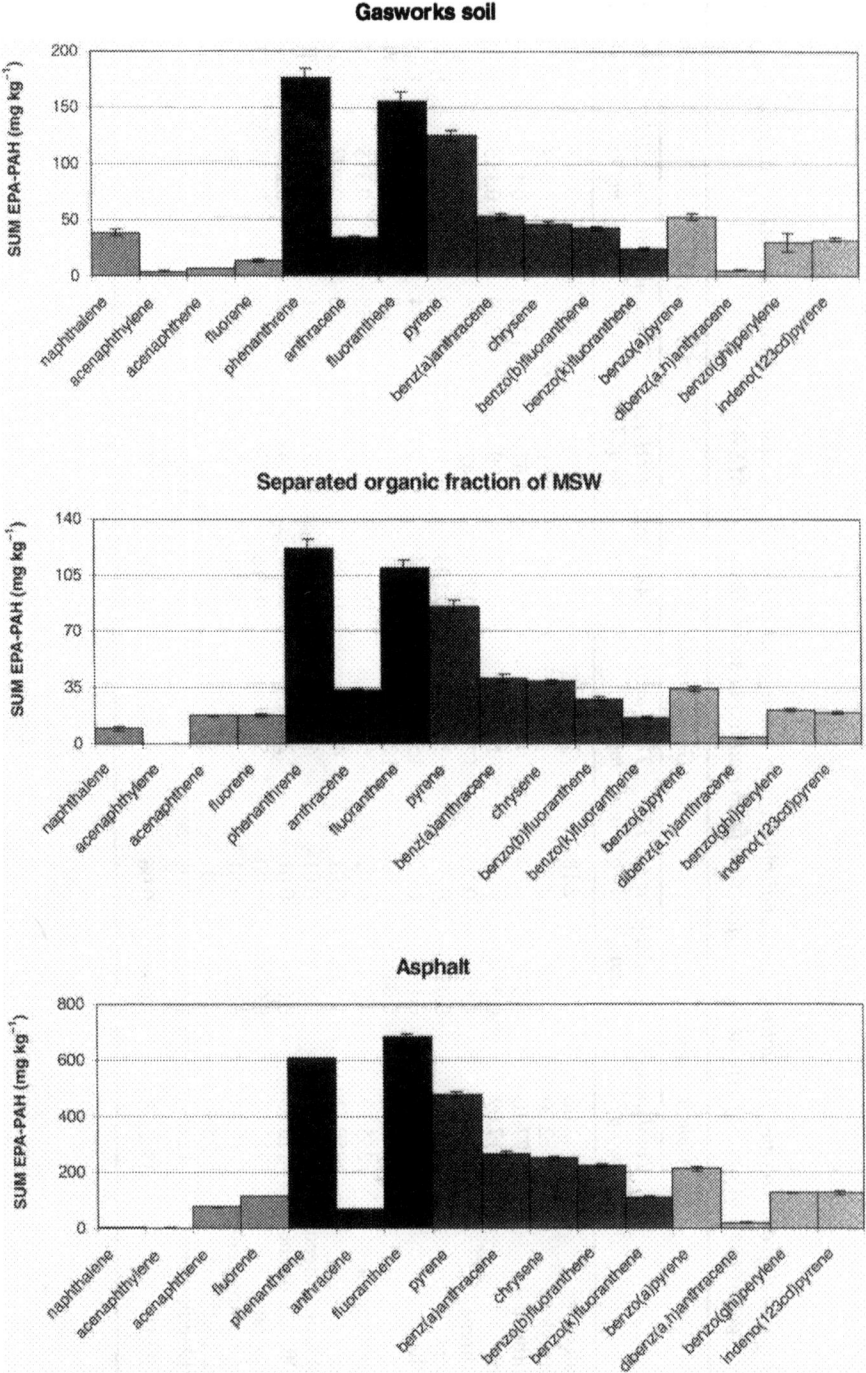

**Figure 2**  *Total content of individual PAHs in gasworks soil (top), separated organic fraction from municipal solid waste (MSW, centre), and asphalt granulate (bottom). Concentrations are measured after Soxhlet extraction. Error bars show average values and standard deviation of duplicate analyses*

**Table 1**  Total (from Soxhlet extraction) and 'available' (from leaching at pH = 12) amounts of individual PAHs in asphalt gamulate, gasworks soil and separated organic fraction from municipal solid waste (OF-MSW)

| PAH | Asphalt gramulate | | | Gasworks soil | | | OF-MSW | | |
|---|---|---|---|---|---|---|---|---|---|
| | Total amount $mg\ kg^{-1}$ | 'Available' at pH 12 $mg\ kg^{-1}$ | % of Total | Total amount $mg\ kg^{-1}$ | 'Available' at pH 12 $mg\ kg^{-1}$ | % of Total | Total amount $mg\ kg^{-1}$ | 'Available' at pH 12 $mg\ kg^{-1}$ | % of Total |
| naphthalene | 3.02 | 0.066 | 2.18 | 38.85 | 1.562 | 4.02 | 9.44 | 0.185 | 1.95 |
| acenaphthylene | 2.30 | 0.018 | 0.78 | 3.94 | 0.060 | 1.51 | – | 0.023 | – |
| acenaphthene | 77.10 | 0.572 | 0.74 | 6.85 | 0.115 | 1.67 | 17.63 | 0.185 | 1.05 |
| fluorene | 117.09 | 0.328 | 0.28 | 13.97 | 0.138 | 0.99 | 17.99 | 0.185 | 1.03 |
| phenanthrene | 610.60 | 0.638 | 0.10 | 176.68 | 0.735 | 0.42 | 122.36 | 1.741 | 1.42 |
| anthracene | 68.24 | 0.050 | 0.07 | 35.34 | 0.103 | 0.29 | 32.99 | 0.410 | 1.24 |
| fluoranthene | 682.02 | 0.155 | 0.02 | 155.88 | 0.194 | 0.12 | 109.84 | 1.252 | 1.14 |
| pyrene | 478.73 | 0.085 | 0.02 | 125.36 | 0.134 | 0.11 | 85.69 | 0.934 | 1.09 |
| benz(a)anthracene | 265.19 | 0.018 | 0.01 | 53.93 | 0.028 | 0.05 | 40.58 | 0.355 | 0.87 |
| chrysene | 250.41 | 0.019 | 0.01 | 46.63 | 0.025 | 0.05 | 38.70 | 0.407 | 1.05 |
| benzo(b)fluoranthene | 224.54 | 0.013 | 0.01 | 42.60 | 0.014 | 0.03 | 27.91 | 0.201 | 0.72 |
| benzo(k)fluoranthene | 113.73 | 0.008 | 0.01 | 24.07 | 0.010 | 0.04 | 16.32 | 0.109 | 0.67 |
| benzo(a)pyrene | 212.47 | 0.011 | 0.00 | 52.61 | 0.016 | 0.03 | 34.33 | 0.216 | 0.63 |
| dibenz(a,h)anthracene | 20.24 | 0.001 | 0.01 | 5.19 | 0.001 | 0.02 | 3.88 | 0.016 | 0.41 |
| benzo(ghi)perylene | 128.96 | 0.010 | 0.01 | 30.07 | 0.012 | 0.04 | 21.03 | 0.076 | 0.36 |
| indeno(123cd)pyrene | 127.94 | 0.008 | 0.01 | 32.98 | 0.011 | 0.03 | 19.73 | 0.074 | 0.37 |
| Σ EPA-PAH | 3382.58 | 2.000 | 0.06 | 844.94 | 3.156 | 0.37 | 598.45 | 6.369 | 1.06 |

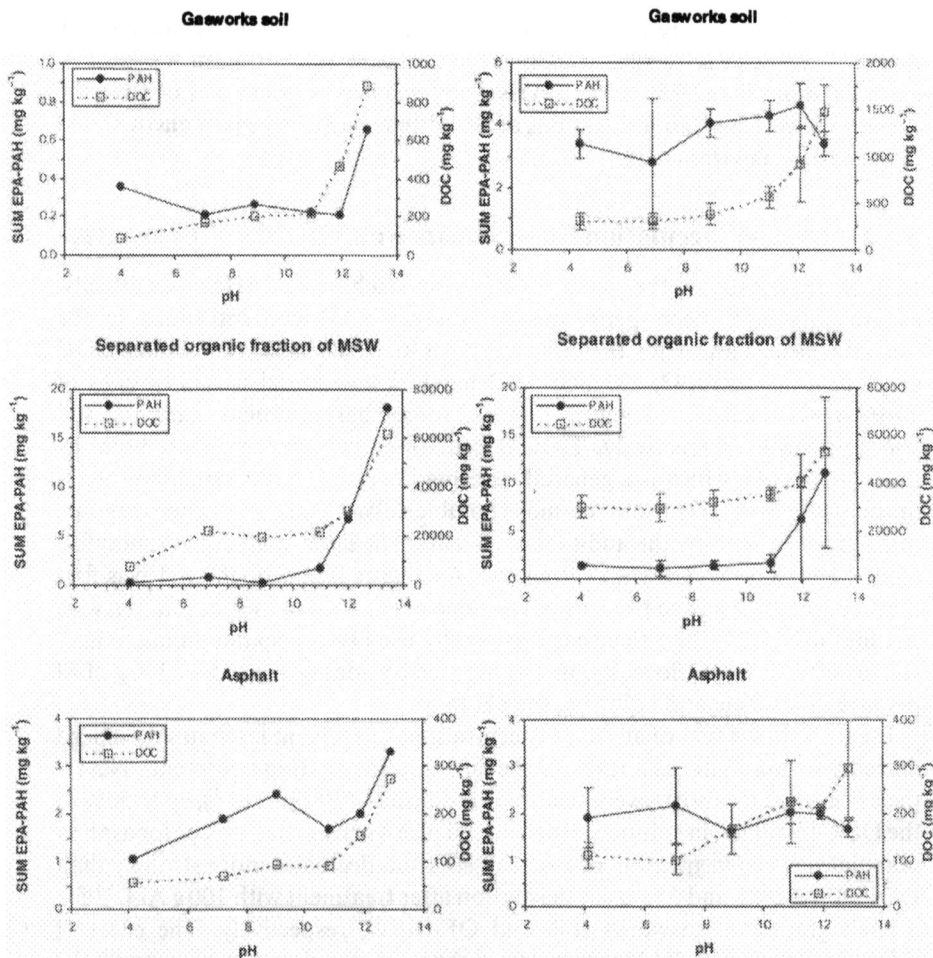

**Figure 3** *Leaching of $\Sigma16$ EPA-PAH and DOC (both in mg kg$^{-1}$ dry solid material) from gasworks soil (top), OF-MSW (centre), and asphalt granulate (bottom) as a function of pH.[2] Initial (single) experiments in which $C_{18}$ solid phase extraction was used for the recovery of PAHs from the leachates are shown on the left. Later experiments (performed in duplicate) in which liquid/liquid dichloromethane extraction was used to recover PAHs from the leachates are shown on the right. Liquid/solid ratio = 10, except for initial experiment with gasworks soil = 2 L kg$^{-1}$. Note the different concentration scales for gasworks soil*

highly alkaline pH values is difficult and likely to cause large interlaboratory differences.

The estimated 'available' amount of PAHs in the soil and waste materials is shown in Table 1. Concentrations of both the individual PAHs and $\Sigma$PAH are shown in units of mg kg$^{-1}$ and percentage of the total amount as measured by Soxhlet extraction. Table 1 clearly shows that the leachable fractions (between

0.06 and 1.06%) are very small relative to the total amounts. These results clearly illustrate the need for a risk-assessment of PAHs in soil/waste on the basis of leaching; more than 99% of the PAHs is not leached even at a pH of 12 in the presence of very high concentrations of DOC that are not normally encountered in the natural environment.

### 6.3.4   DOC-Flocculation Experiments with OF-MSW Leachates

In order to investigate the role of DOC in the leaching of PAHs more directly, experiments have been performed to analyse PAH concentrations in the leachates before and after the removal of DOC by flocculation.[2] Initial DOC concentrations were 43, 131 and 5492 mg $L^{-1}$ and initial concentrations of $\Sigma$EPA-PAH were 47, 159 and 527 $\mu$g $L^{-1}$ for asphalt granulate, gasworks soil and OF-MSW, respectively. DOC-flocculation was accomplished both by lowering the pH, which is a generally accepted method to separate humic acids (insoluble at low pH) from the more soluble fulvic acids and other (smaller) organic acids,[7] and by the addition of Al-salts. The latter method was modified after the method of Laor and Rebhun.[6] As Figure 4 shows, the pH was first lowered from pH = 13 to pH = 6 (*i.e.* the same pH as used in the flocculation with Al), and to pH = 1, the value that is generally used to precipitate humic acids.[7] Independently, DOC flocculation was induced by adding 1 g, 10 g or 100 g of Al per litre of leachate and adjusting the pH to 6.

Figure 4 shows first of all that concentrations of DOC in leachates do indeed decrease substantially as a result of the different flocculation treatments. Nevertheless, even at the most severe treatment (addition of 100 g Al $L^{-1}$), up to 50% of the DOC is still left in solution. However, Figure 4 shows that, except for asphalt granulate, the concentration of PAHs in solution decreases more strongly than DOC; 67%, 28% and 6% is left in solution after treatment with 100 g Al $L^{-1}$ for asphalt granulate, gasworks soil, and OF-MSW, respectively. The effect of DOC-flocculation on PAH removal from the leachates clearly increases with the initial DOC concentration. The data shown in Figure 4 suggest that not all DOC molecules present in the leachates are equally effective in the binding of PAHs and that the most soluble DOC fraction plays only a minor role. This effect is investigated in more detail below by size exclusion chromatography.

---

**Figure 4**   (*opposite*) *Decrease of DOC and $\Sigma$EPA-PAH in alkaline leachates from the gasworks soil (top), separated organic fraction of MSW (centre) and asphalt granulate (bottom) before (pH = 13) and after flocculation of DOC by acidification to pH = 6 and pH = 1, and by addition of 1 g (pH = 6), 10 g (pH = 6) and 100 g Al $L^{-1}$ (pH = 6; 25 g Al $L^{-1}$ for asphalt granulate).[2] Indicated percentages are relative to concentrations of DOC and PAHs in the pH = 13 leachates. Initial DOC concentrations were 43, 131 and 5492 mg $L^{-1}$ and initial concentrations of $\Sigma$EPA-PAH were 47, 159 and 527 $\mu$g $L^{-1}$, for asphalt granulate, gasworks soil, and OF-MSW, respectively*

**Gasworks soil**

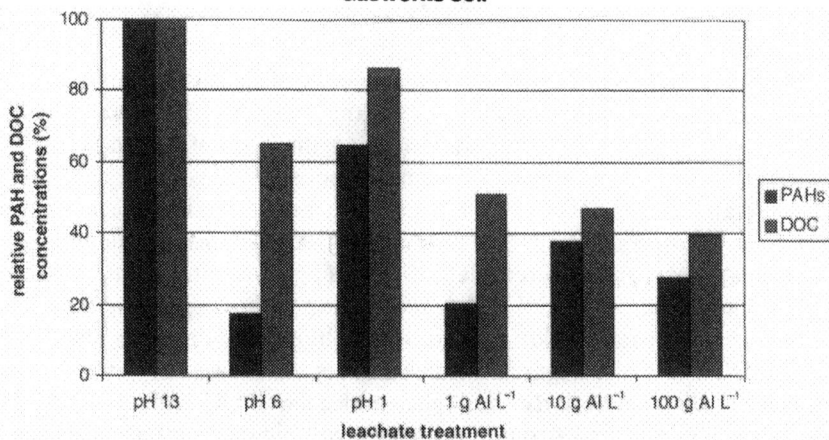

**Separated organic fraction of MSW**

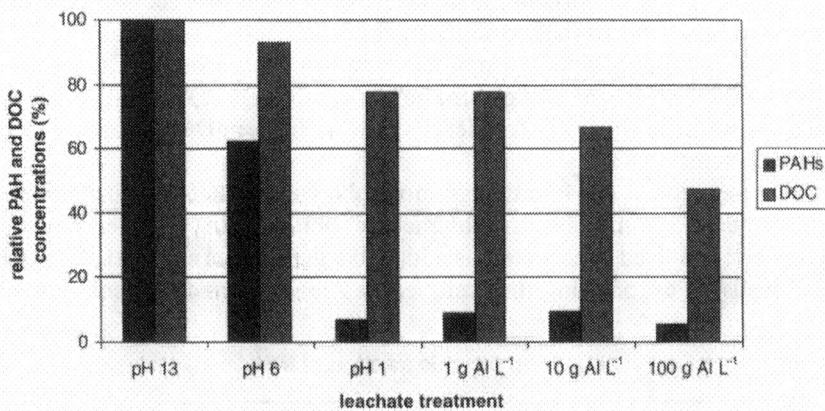

**Asphalt**

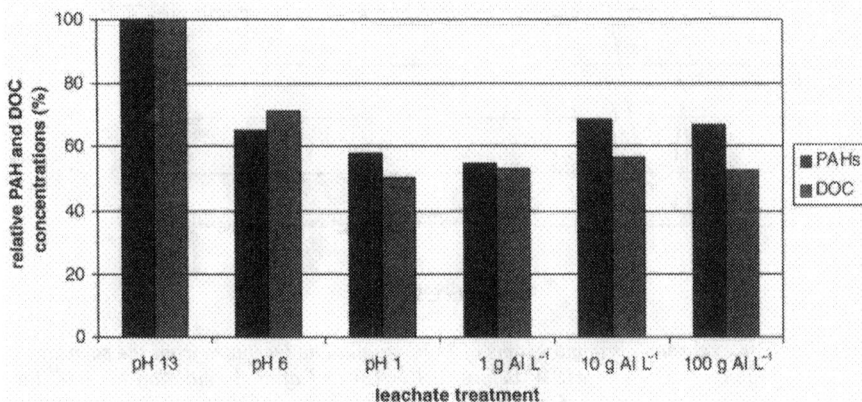

## 6.3.5   Relevant Properties of DOC

Figure 5 shows size exclusion chromatograms of DOC in alkaline leachate from OF-MSW before (pH = 13) and after flocculation of DOC by acidification to pH = 6 and pH = 1, and by addition of 10 and 100 g Al L$^{-1}$.[2] Size exclusion chromatography (SEC) separates molecules on the basis of their size; smaller molecules, that can move through the smaller pores of the column, being more retarded than larger molecules that move only through the macropores. The size spectrum in Figure 5 shows that the DOC in OF-MSW leachates consists largely of molecular material with a retardation time of approximately 12 minutes and higher-molecular size material with a retardation time of approximately 5–10 minutes. The latter retardation times on this column are typical for humic acid size/type DOC (see also Figure 6 below). The results shown in Figure 5 clearly demonstrate that particularly the higher-molecular DOC is removed by the flocculation methods and strongly suggest, in combination with the PAH ana-lyses before and after flocculation (Figure 4), that this fraction is primarily responsible for the binding and solubility enhancement of PAHs in the leachates.

There are various reports in the literature on natural (soil/sediment) organic matter that have shown that particularly the larger and relatively apolar DOC molecules do strongly bind hydrophobic organic contaminants.[8] The results in Figures 4 and 5 are, to the best of our knowledge, among the first to show similar effects for PAHs in waste material leachates with different DOC properties than natural systems.

The association of PAHs with high-molecular weight DOC was further inves-tigated by equilibrating an Aldrich humic acid solution with [14]C-labelled pyrene. [14]C-pyrene dissolved in acetone was added to a glass vessel and coated onto the vessel walls by evaporating the acetone. An Aldrich humic acid solution was

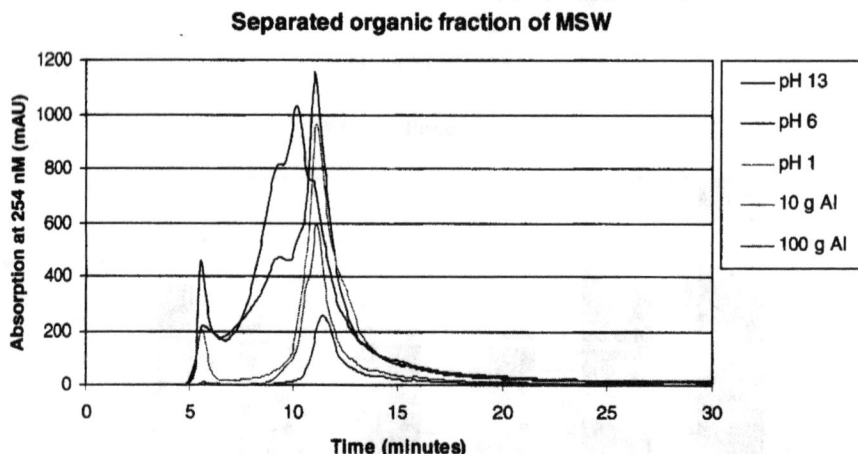

**Separated organic fraction of MSW**

**Figure 5**   *Size exclusion chromatograms of DOC in alkaline leachates from the separated organic fraction of MSW before (pH = 13) and after flocculation of DOC by acidification to pH = 6 and pH = 1, and by addition of 10 and 100 g Al L$^{-1}$.*[2]

prepared by dissolution in nanopure deionised water and purified by removing inorganic mineral matter by high-speed centrifugation. The purified solution was added to the vessels and equilibrated with the $^{14}$C-pyrene that was presumed to be buffered at its aqueous solubility by the coating on the vessel walls (so called 'solubility enhancement' method).[9] The humic acid solution was then separated into different size fractions by running it through the SEC system. These size fractions were sampled and analysed for $^{14}$C-pyrene by liquid scintillation counting. The size fractionation of the humic acid was monitored on-line by both UV (254 nm) absorption and DOC (dissolved organic carbon) analysis.

Figure 6 shows that the pyrene is indeed primarily associated with the higher-molecular size fraction of the humic acid, with retardation times of between approximately 5 and 10 minutes. Very little pyrene is present in low-molecular, including freely-dissolved, form. These results confirm that particularly high-molecular (and generally more hydrophobic) components of DOC can strongly enhance the solubility and, therefore, the leaching of PAHs. This effect seems to be the case not only for natural aquatic systems, but also for waste materials (Figures 4 and 5). The behaviour (*e.g.* solubility and size-fractionation) of DOC is, therefore, a major factor to be considered in both the development and interpretation of leaching tests.

## 6.4 Evaluation of Leaching Processes and Development of 'Availability' Leaching Test

### 6.4.1 Role of DOC in the Leaching of PAHs

We have shown that the leaching of PAHs from the contaminated gasworks soil and the two waste materials increases strongly towards alkaline pH and co-incides with the increase in the solubility of dissolved organic carbon (DOC) in that pH-range (Figure 3). The effect increases with DOC concentration in the leachates and is most pronounced for the separated organic fraction from municipal solid waste (OF-MSW). Therefore, leaching at high pH (pH 12–13) seems appropriate to estimate the fraction of PAHs in the soil/waste matrix that is available for leaching. First estimates of this available fraction, based on leaching at pH = 12 and a liquid/solid ratio of 10 L kg$^{-1}$, are 0.06%, 0.37% and 1.06% for the investigated asphalt granulate, gasworks soil and separated organic MSW fraction, respectively (Table 1). These results clearly illustrate the need for a risk-assessment of PAHs in soil/waste on the basis of leaching; more than 99% of the PAHs are not leached even at a pH of 12 in the presence of high concentrations of DOC that are not normally encountered in the natural environment.

Analyses of PAH concentrations in the leachates before and after the removal of DOC by flocculation clearly show that leached PAHs are predominantly present in a form associated with DOC and corroborate the observed correlation between PAHs and DOC in pH-stat leachates at alkaline pH. Size exclusion chromatography (SEC) of alkaline leachates of the separated organic MSW

**SEC fractionation Aldrich humic acid**

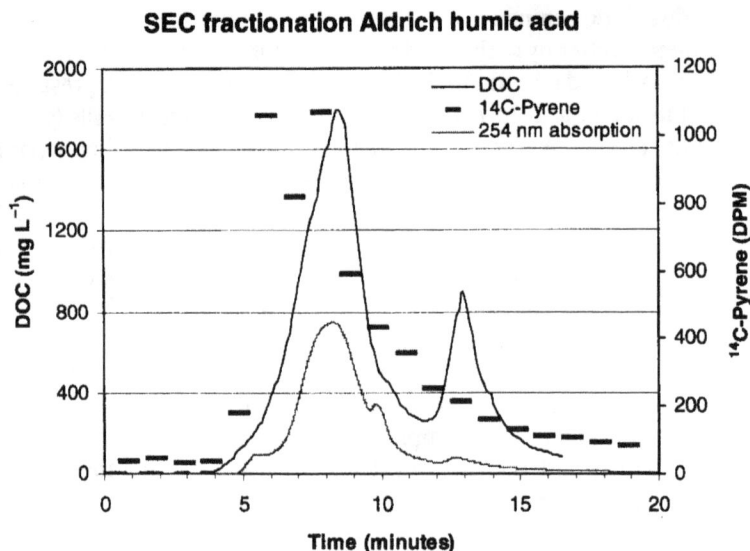

**Figure 6**   *Size exclusion chromatography (SEC) of Aldrich humic acid solution equilibrated with [14]C-pyrene.[2] On-line DOC and 254 nm absorption measurements and [14]C-pyrene analysis of the humic acid size fractions show that the pyrene is primarily associated with the higher-molecular fractions of the humic acid*

fraction has shown that particularly the high-molecular (and probably more hydrophobic) fraction of DOC is responsible for the solubility-enhancement and leaching of PAHs. SEC analyses of a purified humic acid solution equilibrated with [14]C-labelled pyrene have confirmed these findings.

The observed strong solubility-enhancement of PAHs by the solubilisation of DOC from soil and waste materials at high pH can be applied for the development of a standardised 'availability' test that is intended to indicate the maximum amount of the contaminants that can be leached from the soil/waste. We have shown in a recent review that is partly based on results from the European Network on Harmonisation of Leaching/Extraction Tests (see also Chapter 7), that the increase in the solubility of DOC towards high pH, with a plateau at around pH 12–13, is a general property of soils, sediments and many waste materials.[1] These findings justify this approach to the development of a generally-applicable 'availability' test.

Although, as argued above, the increase of DOC concentrations with leachate pH is a general property of soils, sediments and waste materials, the absolute concentrations of DOC at each pH may vary over orders of magnitude, depending on the nature of the soil, sediment or waste material. When an 'availability' leaching test is to indicate the maximum amount of PAHs that can be leached from a material when it is utilised or disposed in the environment, the DOC concentration in the test should reflect the maximum amount that the material can be exposed to in the environment. Therefore, materials such as asphalt granulate, that release relatively small amounts of DOC from their own solid matrix at alkaline pH, should be leached at DOC concentrations that can be encountered when this material is utilised or disposed in the (soil) environment. As recently reviewed,[1] we believe, on the basis of a large number of data on DOC leaching from various soils, sediments and waste materials, that a DOC concentration of 1000 mg L$^{-1}$ is a realistic maximum value that can be encountered in an alkaline waste/soil environment.

## 6.4.2 Development of an 'Availability' Leaching Test

For the above reasons, we propose an 'availability' leaching test where DOC is added to an alkaline (pH 12) leachate at a concentration of 1000 mg L$^{-1}$. Exceptional highly-organic materials such as the separated organic fraction of municipal solid waste (OF-MSW) will under those conditions produce additional DOC, but would probably do the same when utilised/disposed in the (soil) environment. A leachate pH of 12 is chosen because higher values may lead to irreproducible experimental artefacts, which are possibly related to incomplete recovery of PAHs from the leachates. The alkaline pH will prevent added DOC from precipitating/adsorbing to the soil or waste matrix and may also leach additional amounts of DOC from highly-organic wastes such as OF-MSW. We propose Aldrich humic acid as the source of DOC, because it is widely commercially available and has a high binding affinity for hydrophobic organic contaminants such as PAHs.[7]

On the basis of the above framework, we have investigated the influence of leaching time and liquid/solid (L/S) ratio on the 'availability' of PAHs for leaching. Figure 7 shows the effect of the L/S ratio on the measured PAH availability after 48 h of leaching at a pH of 12 and an Aldrich humic acid concentration of 1000 mg L$^{-1}$, for the gasworks soil, separated organic fraction of MSW and asphalt granulate. The availability of PAHs for leaching increases substantially with the liquid/solid ratio. Only at very high L/S ratios of between 500 and 1000 L kg$^{-1}$ the beginning of a plateau in the leaching of PAHs becomes visible. Such high L/S ratios, however, are very impractical when dealing with heterogeneous soils and waste materials because very small amounts of the solid material are leached in a large leachate volume. For example, we used 2.5 g of dry solid in 2.5 L of leachate for the L/S = 1000 L kg$^{-1}$ experiment. Such small quantities of soil/waste material can give rise to a limited reproducibility of the measured availability. Moreover, the use of a 2.5 L Aldrich humic acid solution makes the test very impractical, as the preparation of this solution is time-consuming.[1] Finally, one could argue that such high L/S ratios will not be encountered within relevant time scales in normal waste utilisation or disposal scenarios. For these reasons, we have selected a L/S ratio of 100 L kg$^{-1}$ for the availability test. The available fractions of PAHs in the three soil/waste materials under these conditions are 3.7%, 7.6% and 15.8% for asphalt granulate, OF-MSW and gasworks soil, respectively.

Figure 8 shows the effect of leaching time on the measured PAH availability at

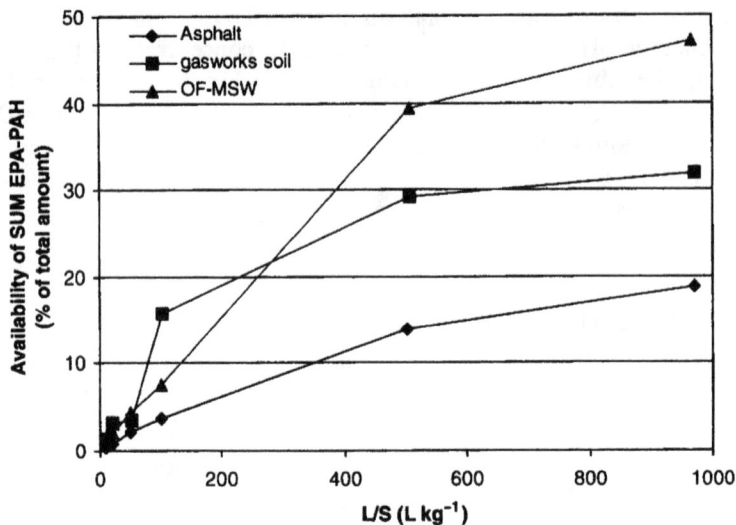

**Figure 7**    *Effect of liquid/solid ratio on the availability (in percentage of total amount) of ΣEPA-PAH for leaching from the gasworks soil, separated organic fraction of MSW and asphalt granulate. Leaching time is 48 h at pH = 12 and 1000 mg $L^{-1}$ Aldrich humic acid. The total amounts of PAHs are shown in Table 1*

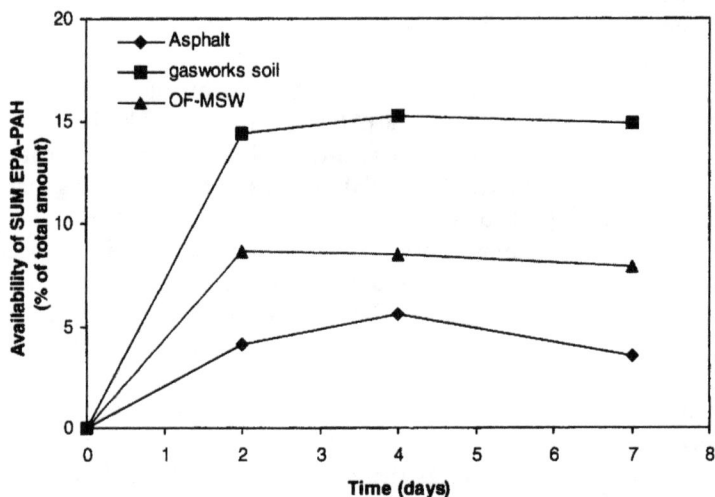

**Figure 8**    *Effect of leaching time on the availability (in percentage of total amount) of ΣEPA-PAH for leaching from the gasworks soil, separated organic fraction of MSW and asphalt granulate. Liquid/solid ratio = 100 L $kg^{-1}$, pH = 12 and 1000 mg $L^{-1}$ Aldrich humic acid. The total amounts of PAHs are shown in Table 1*

a L/S ratio of 100 L kg$^{-1}$, a pH of 12 and an Aldrich humic acid concentration of 1000 mg L$^{-1}$, for the gasworks soil, separated organic fraction of MSW and asphalt granulate. Except for the asphalt granulate, the changes in leachate PAH concentrations after 48 h reaction time are relatively small. The cause of the decrease in PAH leaching from asphalt granulate after 4 days is unclear. Based on these findings and practical considerations, we have selected a leaching time of 48 h for the availability test. This choice enables that the entire cycle of *leaching test → extraction of PAHs → analysis of PAHs* can be completed within one week. The available fractions of PAHs in the three soil/waste materials under these conditions are 4.1%, 8.7%, and 14.4% for asphalt granulate, OF-MSW and gasworks soil, respectively. It should be noted that the values measured in the two separate experiments shown in Figure 7 and 8, for the conditions that have been selected for the final 'availability' test, deviate by less than 15%. These results obtained for three very different soil/waste materials suggest that the reproducibility of the proposed 'availability' test is very good. A limited round robin test of this procedure among six different laboratories has confirmed these findings.[1]

The general features that have been observed in the leaching of PAHs from the three very different soil and waste materials that we have studied, particularly the key role of DOC therein, the additional experiments on the effect of leaching time and liquid/solid ratio, and practical considerations, have lead us to define a batch availability leaching test based on the following conditions:

- leachate of 1000 mg L$^{-1}$ Aldrich humic acid
- pH = 12
- liquid/solid ratio = 100 L kg$^{-1}$
- leaching time = 48 h
- leachate separation by centrifugation

Centrifugation rather than filtration is proposed because of the difficulty in filtering the highly humic leachates, and to prevent losses of PAHs by adsorption to filtration membranes/equipment and/or by volatilisation. The full procedure is described in detail eleswhere.[1]

## 6.5   Conclusions

### 6.5.1   Leaching Processes of PAHs

Very similar features have been observed in the leaching of PAHs from three very different soil and waste materials: a gasworks soil, the mechanically-biologically separated organic-rich fraction of municipal solid waste (OF-MSW) and a tar-containing asphalt granulate. The leaching of PAHs increases strongly towards alkaline pH and coincides with the increase in the solubility of dissolved organic carbon (DOC) in that pH-range. The effect increases with DOC concentration in the leachates and is most pronounced for the separated organic fraction from municipal solid waste. At pH = 12 and a liquid/solid ratio of 10 L kg$^{-1}$, only 0.06%, 0.37% and 1.06% of the total amount of PAHs was leached from the investigated asphalt granulate, gasworks soil and OF-MSW, respectively. These results clearly illustrate the need for a risk-assessment of PAHs in soil/waste on the basis of leaching.

Analyses of PAH concentrations in the leachates before and after the removal of DOC by flocculation have clearly shown that leached PAHs are predominantly present in a form associated with DOC and corroborate the similar features in the leaching of PAHs and DOC as a function of pH. Size exclusion chromatography (SEC) of alkaline leachates of the separated organic MSW fraction has shown that particularly the high-molecular (and probably more hydrophobic) fraction of DOC is responsible for the solubility-enhancement and leaching of PAHs. SEC analyses of a purified humic acid solution equilibrated with $^{14}$C-labelled pyrene have confirmed these findings.

It has been shown earlier that the increase in the solubility of DOC towards high pH, with a plateau at around pH 12–13, is a general property of soils, sediments and many waste materials. Therefore, the strong solubility-enhancement of PAHs by the solubilisation of DOC from soil and waste materials at high pH, demonstrated in this study, has been used as the basis for the development of a generally-applicable 'availability' leaching test that is intended to indicate the maximum amount of contaminants that can be leached from the soil/waste.

### 6.5.2 'Availability' Leaching Test

The enhancement of PAH leaching at alkaline pH and the estimated available PAH fractions have been found to increase with the DOC concentration in the leachates. Although the increase of DOC concentrations with leachate pH is a general property of soils, sediments and waste materials, the absolute concentrations of DOC at each pH may vary over orders of magnitude, depending on the nature of the soil, sediment or waste material. An 'availability' leaching test is meant to indicate the maximum amount of PAHs that can be leached from a material when it is utilised or disposed in the environment. Therefore, the DOC concentration in the test should reflect the maximum amount that the material can be exposed to in the environment. For these reasons an availability leaching test is proposed based on an alkaline leachate of pH = 12, to which DOC is added in the form of a generally available (Aldrich) humic acid at a concentration of 1000 mg L$^{-1}$. Additional experiments and practical considerations have led to the selection of a liquid/solid ratio of 100 L kg$^{-1}$ and a leaching time of 48 h. The available fractions of PAHs in the three soil/waste materials under these conditions, based on two separate experiments, are $3.9 \pm 0.3\%$, $8.1 \pm 0.7\%$ and $15.1 \pm 1.0\%$ for asphalt granulate, OF-MSW and gasworks soil, respectively. These results obtained for three very different soil/waste materials suggest that the reproducibility of the proposed 'availability' test is very good (*i.e.* within 15%). A limited round robin test of this procedure among six different laboratories has confirmed these findings.[1]

### 6.6   Acknowledgements

The work described in this chapter was partly funded by two EU projects,

SMT-4-CT97-2160 on the development of leaching tests for organic contaminants, and EVK1-CT-1999-00029 on groundwater risk assessment at contaminated sites (GRACOS).

## 6.7  References

1. R.N.J. Comans (ed.) *Development of Standard Leaching Tests for Organic Pollutants in
   Soils, Sediments and Granular Waste Materials*, Final Report, EU project SMT-4-
   CT97-2160, European Commission, Brussels, Belgium.
2. G.D. Roskam, L.M. Shor and R.N.J. Comans, *Role of Dissolved Organic Carbon in the
   Leaching of Polycyclic Organic Hydrocarbons (PAHs) from Soil and Waste Materials*,
   submitted for publication.
3. GRACOS, Groundwater Risk Assessment at Contaminated Sites (GRACOS), *1st
   Annual Progress Report* (contract EVK1-CT-1999-00029), University of Tübingen,
   Germany.
4. R.P. Schwarzenbach, P.M. Gschwend and D.M. Imboden, *Environmental Organic
   Chemistry*, John Wiley & Sons, New York, 1993, 681.
5. Y.-P. Chin, W.J. Weber and B.J. Eadie, *Environ. Sci. Technol.*, 1990, **24**, 837.
6. Y. Laor and M. Rebhun, *Environ. Sci. Technol.*, 1997, **31**, 3558.
7. F.J. Stevenson, *Humus Chemistry*, 2nd edition, John Wiley & Sons, New York, 1994,
   496.
8. R.G. Luthy *et al.*, *Environ. Sci. Technol.*, 1997, **31**, 3341.
9. K.M. Danielsen, Y.-P. Chin, J.S. Buterbaugh, T.L. Gustafson and S.J. Traina, *Environ.
   Sci. Technol.*, 1995, **29**, 2162.

CHAPTER 7

# Harmonisation of Leaching/Extraction Procedures for Sludge, Compost, Soil and Sediment Analyses

## H.A. VAN DER SLOOT

Netherlands Energy Research Centre (ECN), Petten, The Netherlands

## 7.1 Introduction

In the framework of the Standards, Measurement & Testing programme (European Commission), a Network for the Harmonisation of Leaching/Extraction Tests was initiated in 1995.[1] The background for starting this network was the increased use of leaching test methods in different areas – waste treatment and disposal; incineration of waste; burning of waste fuels; soil clean-up and reuse of cleaned soil; sludge treatment; use of compost from different sources; and use of secondary raw materials in construction.[2-9] In the first phase of the Network, a first step has been made towards harmonisation of leaching/extraction tests by bringing together experts and discussing differences and similarities in testing between different fields. The direct result from this activity is a book published in 1997 on 'Harmonisation of Leaching/Extraction Tests'.[10] As a follow up the project 'Technical Work in Support of the Network' (SMT4-CT96-2066)[11,12] and the project 'Leaching of Organic Contaminants' (SMT4-CT97-2160)[12-14] were initiated. The Network on Harmonisation of Leaching/Extraction Tests received further support from the European Commission to organise expert meetings on specialised topics – leaching of organic contaminants, modelling of leaching, lab field relationships. In addition, intensive interactions at CEN and ISO level have been taken on to broaden the experience in using more elaborate methods of leaching to assess material behaviour.

Although important work has been carried out in the field of waste and the field of construction materials, this chapter focuses on material containing organic matter. These materials have a common characteristic – namely the role of particulate organic matter and dissolved organic matter (DOC), which is has proven to be a key factor in the release of both inorganic and organic contaminants. In this chapter, the following aspects will be addressed: scenario approach *versus* arbitrary approaches of assessment, total composition *versus* leaching behaviour, release controlling factors (*e.g.* pH, DOC, redox), test methods and recent developments, modelling of release, and regulatory developments.

## 7.2   Scenario Approach *versus* Arbitrary Approaches of Assessment

Recent developments in environmental impact assessment focus on long term release from materials taking as much as possible material behaviour and external influences on material behaviour into account.[15-20] The total composition is no longer a suitable parameter as in many cases this will lead to an over-conservative estimate. In instances where leaching has been applied as a basis for regulatory control, simple methods[21-24] have been promoted that address only a very limited scenario. The EPTOX, later developed into the TCLP test, focused at a co-disposal scenario of waste with municipal solid waste is an example of such an approach.[21] The primary development of DIN 38414 S4 was focused on the release from sediments, which by the nature of the materials tested was focused at leaching at neutral pH values.[22] In both cases the tests were taken outside their context and used for totally different purposes. The TCLP has been applied in a general manner to test waste for disposal, in which the test was disconnected from its primary aim. This has led to unjustified acceptance of wastes for disposal in monofill situations.[25] In the case of DEV S4, alkaline material may show leaching behaviour, which would lead to unjust acceptance or rejection under conditions where the material undergoes long term changes in leaching properties. These misuses of the tests are a result of their design for a narrower scope. The increased understanding of the complexity of factors playing a role in assessing long term behaviour of materials has led to the development and standardisation of tests that reflect more closely the actual behaviour in practice. Release under conditions that mimic the practice as closely as possible is the main basis for assessment, then the emphasis will shift to development of methods that provide information on mechanistic aspects of leaching. In CEN/TC 292, characterisation of waste, this has led to development of a guideline ENV 12920,[26] in which waste-specific properties and scenario conditions are taken into account. The methodology contains several steps, some of which are using chemical, biological, physical and leaching test methods. A selection of tests is to be made depending on the question(s) to be answered, the waste under consideration and the scenario to be evaluated. The tests developed in CEN TC 292, as a result of this change, comprise a percolation test,[27] a pH dependence leaching test,[28,29] and recently, a dynamic method for release from monolithic materials.[30] In the context of the development of criteria and methods as specified in Annex II of the EU Landfill Directive,[31] this scenario approach has shown very promising possibilities.[18] The tests developed in CEN TC 292 are based on a hierarchy in testing – characterisation, compliance and on-site verification. This approach is equally applicable to other fields outside the waste area. Therefore, a similar basis of testing and using a scenario type of approach is suggested for the fields of sludge, compost, soil and sediment evaluation as well. The proper question to be asked is what information a test should provide and how the results will ultimately be used in the context of daily control or in a regulatory framework.

A range of questions can be asked in relation to sludge amendment, compost application, (contaminated) soil use, incineration of biowaste and disposal of all of these materials in case the quality is not acceptable for normal utilisation. Such as: how can a limit for repeated application of sewage sludge or compost be derived based on accumulation of contaminants in soil? What method will provide a good measure for uptake in plants? How to assess the long-term release from contaminated soil *in situ*? What is the release for contaminants from sediments undisturbed at the bottom of a waterway, when applied on land or disposed off? In the latter case the main material properties do not change dramaticallly, but the exposure conditions of the same material are changing very significantly. Are different tests needed for these different exposure conditions or can the same test be used to assess different conditions? How to evaluate treatment methods to improve the environmental quality of sludge, compost, soil or sediments?

All of these questions require an evaluation of factors influencing the release of constituents and the main release mechanisms involved. The test methods currently standardised[27–30] provide such answers. In the framework of the Network Harmonisation of Leaching Extraction Tests, the importance of pH, redox and inorganic and organic complexants such as dissolved organic matter was already identified.[10,11] For the materials discussed in the context of this chapter, the main release-controlling factor is the role of particulate and dissolved organic matter (DOC). Details of which are addressed in Section 7.4 in relation to mobilisation of DOC and in Section 7.6 in relation to modelling metal–DOC interactions.

In relation to different types of judgement for the same material, the pH dependence test provides a basis as illustrated in Figure 1. The leaching characteristic for Cd from this heavily sewage sludge-amended soil illustrates the Cd leachability under different exposure conditions, which may be relevant for the material. In case of inhalation/ingestion, the low pH domain is the most relevant condition to judge possible uptake. When exposed to acidic conditions, for instance by an overlying soil with a more acidic character, the leachability is very sensitive to small pH changes. The effects of treatments such as liming or cement stabilisation (occasionally applied to reduce release) can be derived from the pH dependence test data. Cement stabilisation is not favourable for Cd due to the induced mobilisation of DOC with increasing pH, which can facilitate the mobilisation of Cd as illustrated by the pH dependence test data in the pH range pH > 9. These order of magnitude changes as a result of the pH are important to be realised by a user of test data.

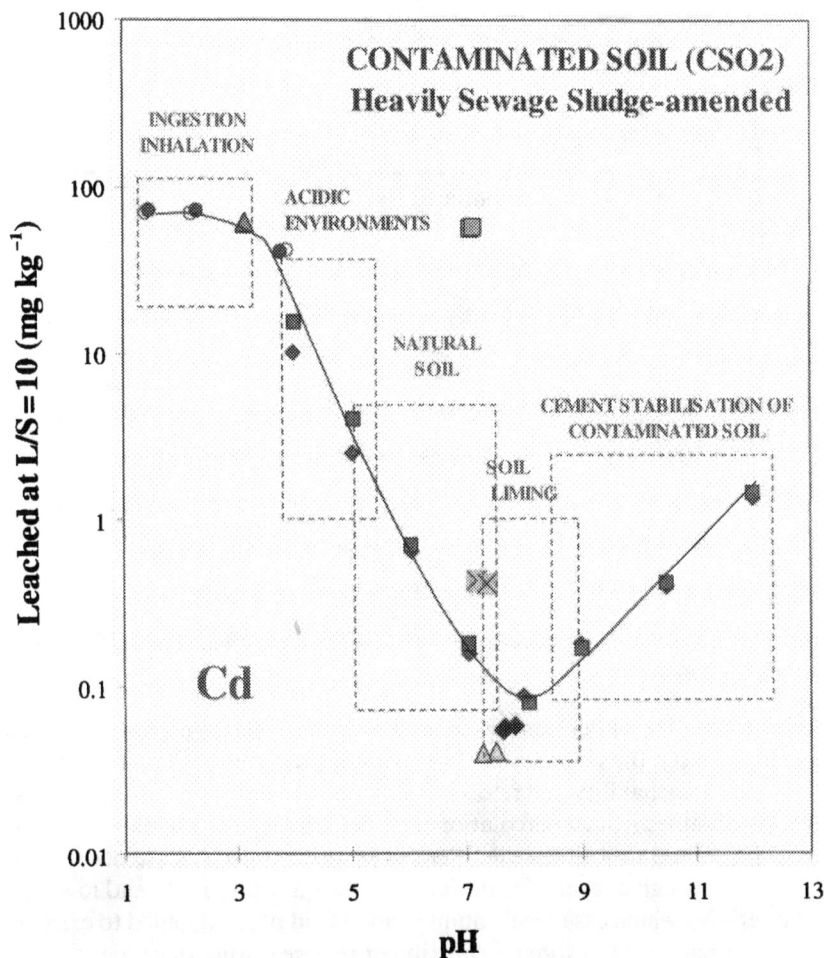

**Figure 1**  *Uses of pH dependence test data in different decisions in relation to soil and soil uses. Material: heavily sewage sludge-amended soil CSO2*[12]

## 7.3   Total Composition *versus* Leaching Behaviour

In many regulations, particularly in the fields of sludge, compost, soil and sediments, total composition or a partial dissolution is often used as a basis of reference. From an impact and long-term environmental behaviour point of view also for these matrices, an approach based on leaching would be more suitable. In a meeting organised in relation to the preparation of the Network Harmonisation of Leaching/Extraction Tests held in Maastricht in 1997 the issue of total composition *versus* leaching was addressed.[32] It was stated that different levels could be distinguished, which each having their own role in the overall assessment – total, available or potentially leachable and actual leaching under specified conditions. In Figure 2 a graphical presentation is given of the different

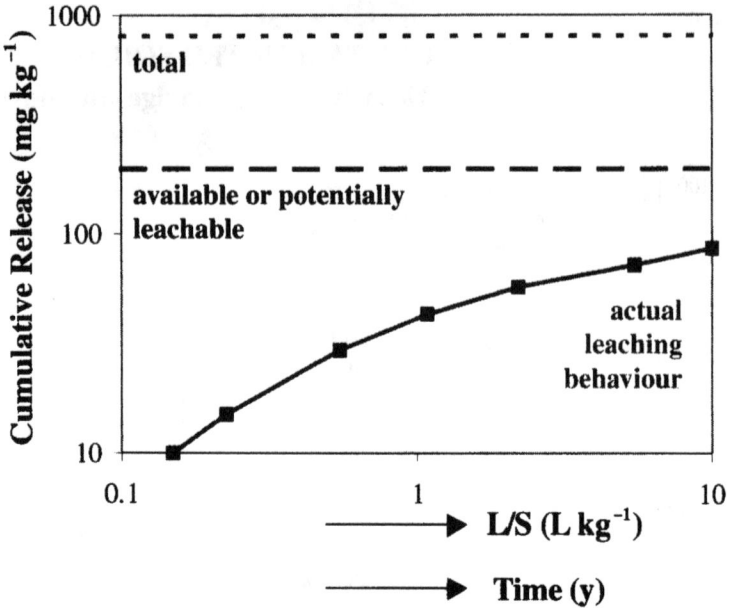

**Figure 2** *Relationship between total, potentially leachable and actual leaching from 'granular' materials*

levels of release. Dependent on the constituent of interest, significant differences may exist between the three levels. In this case, total composition is compared with potential leachability and release as a function of liquid to solid ratio (L/S in $L \, kg^{-1}$) as obtained from a percolation test. The results obtained as a function of L/S can be related to a time scale. Percolation-controlled release occurs when water flows through a layer of material with low infiltration rate and low liquid to solid ratio, in which case local equilibrium at field pH is assumed to exist. The information required to estimate constituent release during this scenario is the field geometry, the field density, the anticipated infiltration rate, the anticipated field pH, the anticipated site-specific liquid to solid ratio, and the constituent solubility at the anticipated field pH. The anticipated site-specific liquid to solid (L/S) ratio represents the cumulative liquid to solid ratio that can be expected to contact the material layer over the estimated time period. It is based on the infiltration rate, the contact time, the fill density and the fill geometry and can be determined according to Hjelmar[33] and Kosson *et al.*:[16]

$$LS_{Site} = 1000 \times \frac{(inf) \times t_{year}}{\rho H_{Fill}}$$

where $LS_{Site}$ = anticipated site-specific liquid to solid ratio [$L \, kg^{-1}$]; inf = anticipated infiltration rate [$cm \, year^{-1}$]; $t_{year}$ = estimated time period [year]; $\rho$ = material density [$kg \, m^{-3}$]; and $H_{fill}$ = material depth [m].

Over an interval of 100 years or longer, L/S values greater than 10 may be obtained for cases that have relatively high rates of infiltration or limited placement depth.

An illustration of the factors controlling release in different matrices is given in Figure 3. This schematic for material behaviour is generally applicable for different matrices. It is clear that, as a function of pH, order of magnitude differences in actual leachability can occur.

In terms of total composition, distinction must be made between different acid mixtures. In fact only upon complete dissolution using HF can a true total be obtained. In Figure 4 a comparison is given of three types of 'total' composition analysis – *aqua regia*, $HNO_3$ in a pressure bomb, $HNO_3/HClO_4$ and *aqua regia* + HF – with availability according to NEN 7341[35] for four materials. The materials are: a soil contaminated by heavy amendment with sewage sludge, a heavily contaminated river sediment, sewage sludge from an installation receiving rural community waste and compost obtained from source-separated putricible fraction from municipal waste. It illustrates a significant difference for many elements between total composition and the maximum leachable quantity under rather extreme conditions unlikely to occur in practice. Partial dissolution methods such as *aqua regia* and nitric acid digestion may lead for some elements (*e.g.* Pb, Zn and Ni) to rather comparable results. For other elements, however, significant differences may occur (*e.g.* As, Cr and $PO_4$ as P).

If also for total composition analysis a distinction is made in characterisation and compliance level, one could state that a true total would be the characterisa-

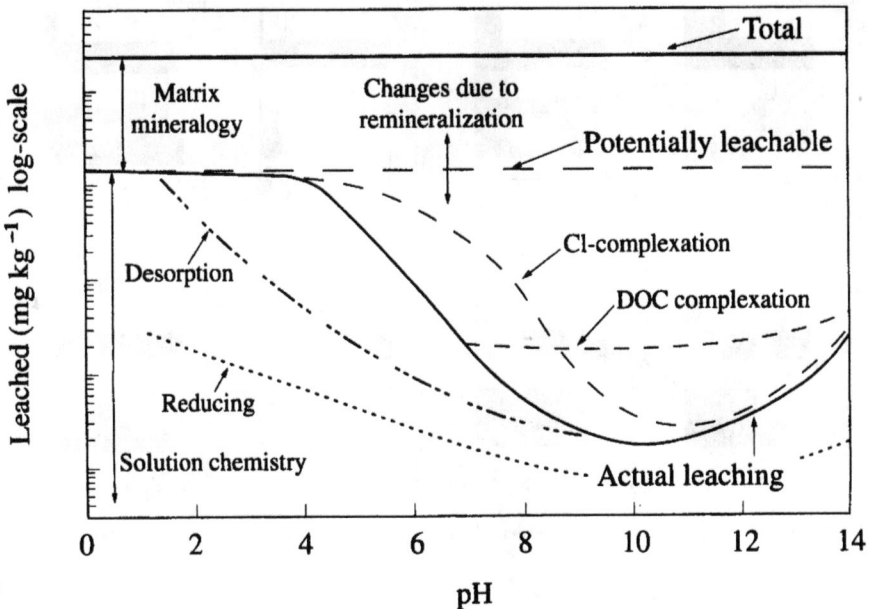

**Figure 3** *Factors controlling release from a material as a function of pH*[10]

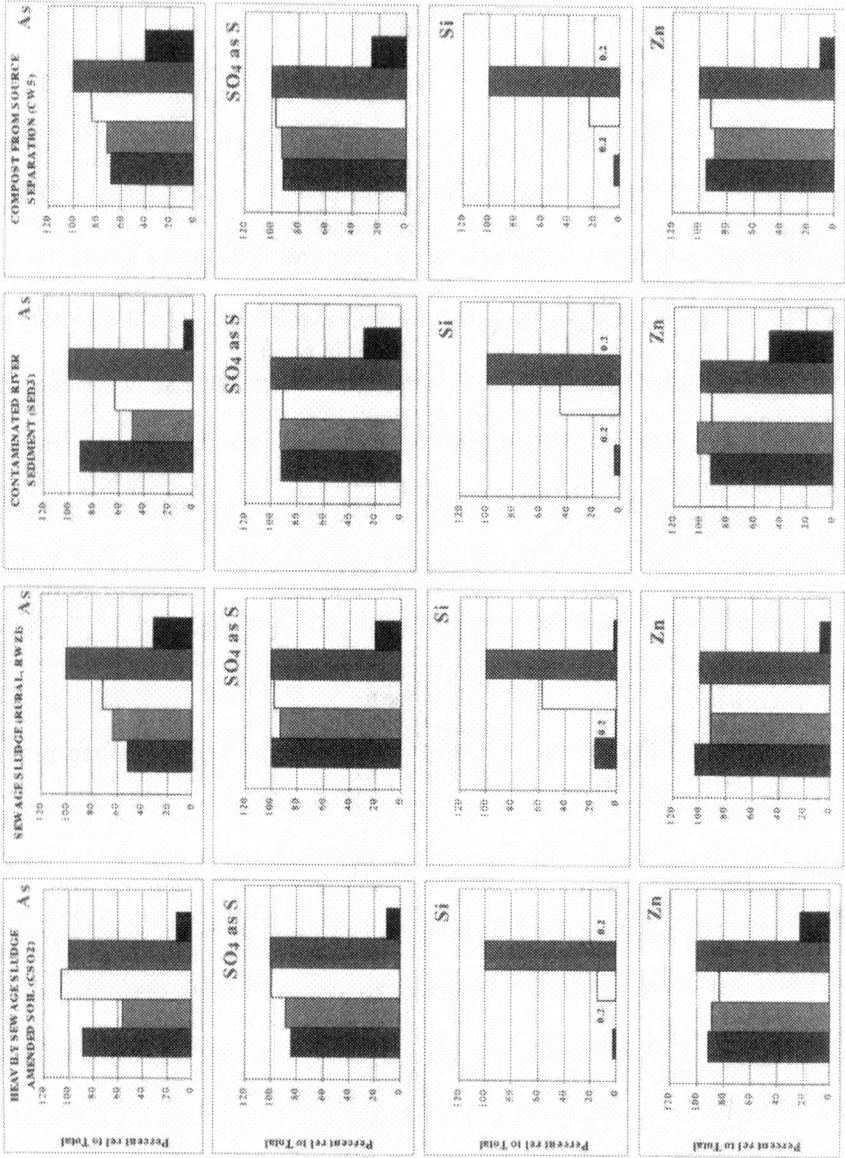

**Figure 4**   *Comparison of total – aqua regia, $HNO_3$, $HNO_3/HClO_4$, aqua regia + HF – with NEN 7341.*

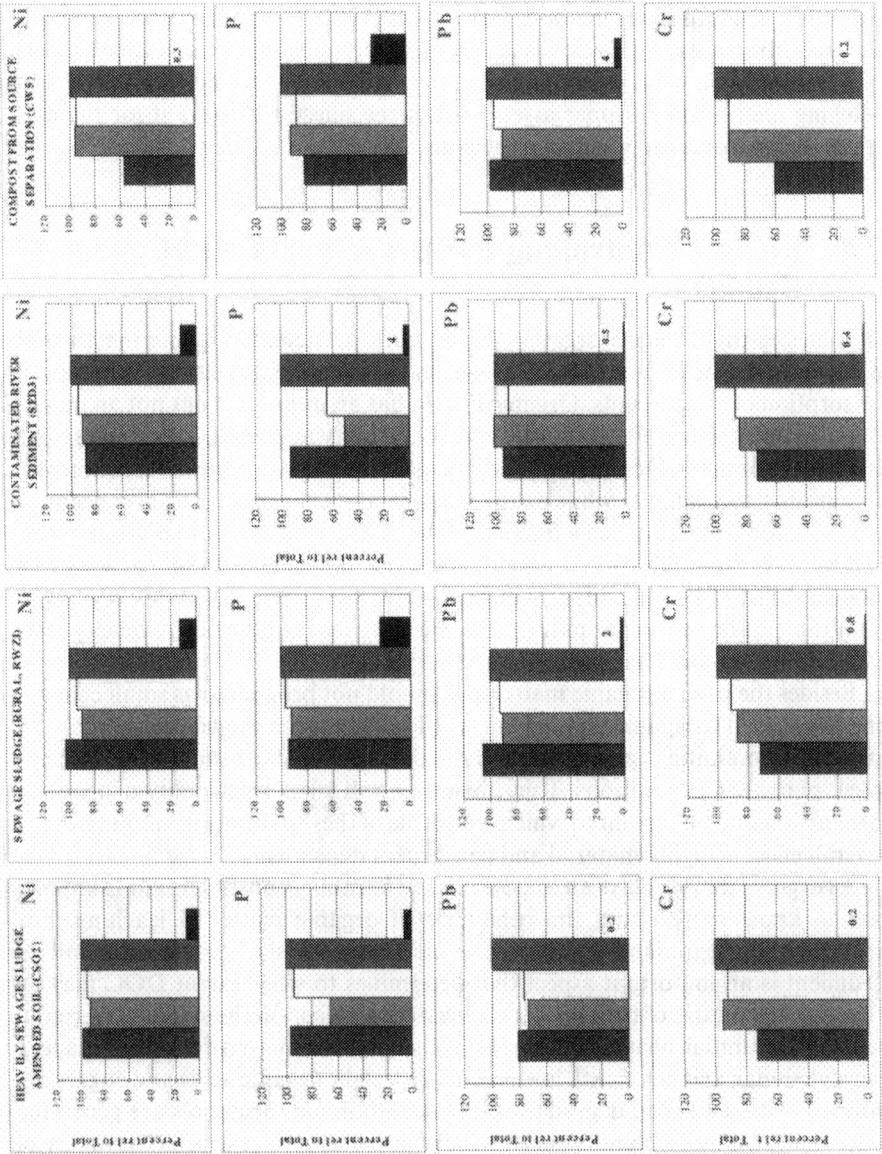

**Figure 4** (continued)

tion level, as only a true total obtained after complete dissolution of the matrix can be called 'total'. Partial dissolution methods, which are easier to perform and provide consistent results for a specific matrix can then be used for practical purposes as compliance method. For an evaluation of environmental impact, however, the total composition is too far away from reality to be suitable as basis of reference for acceptance. Therefore, it is important to adopt and/or develop leaching methods for proper assessment of the questions relevant for the matrices sludge, compost, soil and sediment.

## 7.4 Release Controlling Factors (*e.g.* pH, DOC, Redox)

In matrices that contain organic matter, both particulate organic matter and dissolved organic matter play a key role in the potential for either mobilisation or sorption of constituents. Organic matter has an important but not an unambiguous influence on the concentration, availability and transport of metal ions in the environment. On the one hand organic matter lowers the metal concentration by its strong affinity while on the other hand it enhances the total concentration of soluble metals when a fraction of the organic matter is dissolved in water (DOC). It is the soluble fraction of the organic matter (DOC) that can lead to a high mobility of the metal ions while the metal ion activity is low to very low in solution. Characteristic for natural organic matter is the high heterogeneity of its affinity for ions and its complex composition.

Besides the role of organic matter, pH should not be ignored as small changes in pH can lead to significant differences in leachability as shown before.[10,11,6] In matrices containing organic matter, biological activity can result in the development of reducing conditions. This, obviously, will affect leachability of a variety of constituents. The extent to which either one of these parameters dominates the overall effect is highly material and constituent dependent.

As organic matter plays an important role for many constituents, it is important to know more about the behaviour of organic matter in leaching. For instance, the degradability of the organic matter in sludge, compost, soil or sediment is an important aspect that determines to what extent DOC may be formed. The nature of the DOC is also very dependent on the status of degradation. In the initial phases of degradation of organic matter-rich materials (*e.g.* sewage sludge and compost), low molecular weight organic acids and sugars are formed. After extended degradation, fulvic and humic substances are generated with significantly different properties in terms of metal interaction. The nature of the DOC will thus differ between matrices containing organic matter both in level and in type. Clearly, this important subdivision of DOC implies that generic parameters such as loss on ignition (LOI) and total organic carbon (TOC) analysed in the solid are not sufficiently discriminating in identifying truly degradable organic matter.[34] Pure carbon and plastics are partially included in a TOC analysis. The analysis of DOC ( = TOC in water) is considered to be more discriminating, as the DOC measured at different pH values can also discriminate between types of DOC. At neutral pH, only low molecular weight DOC can be mobilised. At high pH, high molecular weight organic matter, such as humic substances, is mobilised. Particularly, the latter can act as carrier for both metals and organic contaminants.

In the framework of the technical work in support of the Network Harmonisation Leaching/Extraction tests,[11] DOC leachability as a function of pH has been measured for materials with different levels of organic matter. In Figure 5 the DOC as a function of pH for natural soil, contaminated soil, sewage sludge, compost and sediment is given. It shows that sewage sludge is still a highly reactive material with a very high DOC level even at neutral pH. In natural soil and also in mildly contaminated soil degradation has been active for a long time and a level of residual degradation is observed in which only limited amounts of low molecular materials are formed. This is an indication of the level of degradation reached. The measurement of DOC at neutral pH has been proposed as a

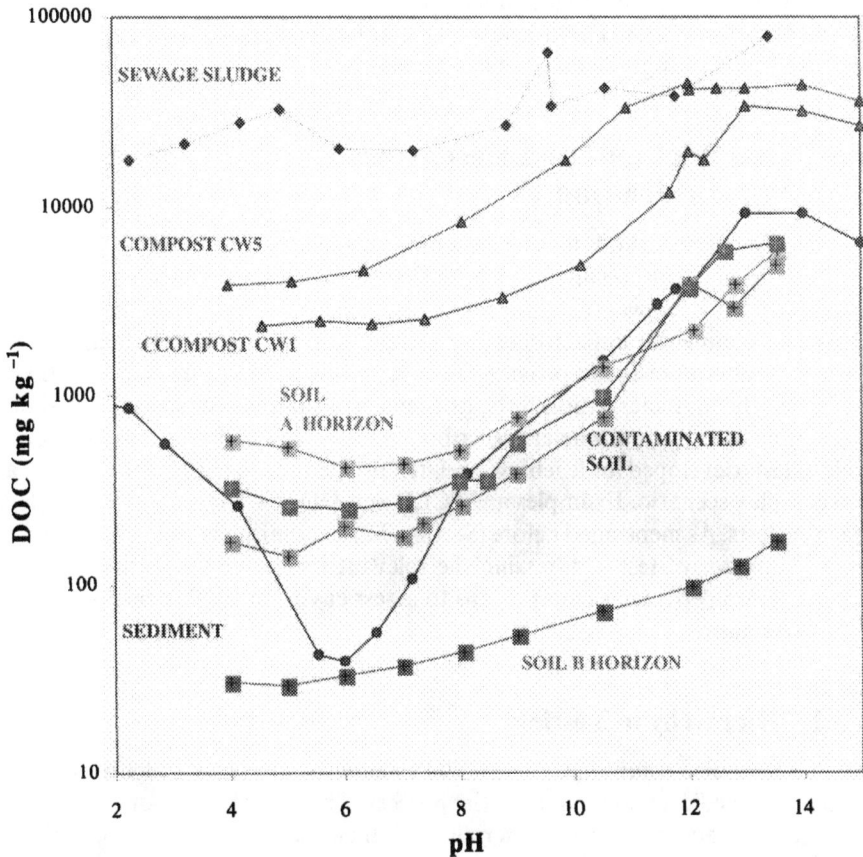

**Figure 5** *Relationships between DOC in a wide range of materials indicating different stages of degradation of (natural) organic matter. Indications of different nature of DOC released in different pH ranges (SMT4-CT96-2066)[11]*

means of assessing the status of degradation of a material.[36] Comparable properties have recently been suggested.[37] In this case sewage sludge is more reactive than compost, which in turn is more reactive than soil. In sediment, it appears that there is a gap in the DOC between pH 5 and 7. The lack of material in this range might be explained by the fact that low molecular weight DOC once generated diffuses easily out of sediments and thus sediment is depleted of the low molecular weight fraction. This mechanism still needs to be verified in other sediment samples. Work on the highly degradable fraction of municipal solid waste has shown that the final stage of degradation of organic rich material points at a strong similarity of DOC leaching with that of natural soil.[36]

The emphasis on DOC measurement and its further fractionation[38] is important to be able to quantify effects on metal and organic contaminant mobility. The information on DOC as a function of pH in conjunction with information on site densities of particulate and dissolved organic matter provides the basis for modelling such interaction (see Section 7.6). The nature of the organic matter from sewage sludge, compost and soil may differ in terms of site densities and nature of sorption sites. This aspect is as yet not sufficiently covered as only data for peat[39] and a range of soils[40] are available.

## 7.5  Test Methods and Recent Developments in Standardisation

Standardised test methods for evaluation of material behaviour are important to produce comparable results in testing. World-wide many leaching tests are available.[10,41,42] However, to address release of constituents from materials to assess environmental impact does not require a multitude of test methods. A limited number of tests can provide the appropriate information. Such methods are generally characterisation tests, as they aim at parameter specific relationships or mechanistic parameters describing release. The problem with the multitude of tests developed for leaching is related to attempts to mimic with one test a too complex scenario. Examples are TCLP and DIN S4 and similar single step leaching tests. As mentioned before (Section 7.2) a scenario approach involving a limited number of tests addressing the relevant issues for the scenario under consideration provides a better means to assess environmental impact and long term behaviour.

### 7.5.1  Hierarchy in Testing

As a result of discussions in CEN/TC 292 a hierarchy of tests was identified, each with their specific function in the entire process of material behaviour evaluation. Tests to characterise waste materials and their behaviour can generally be divided into three categories:
1. 'Basic Characterisation' tests are used to obtain information on the short and long term leaching behaviour and characteristic properties of waste materials. Liquid/solid (L/S) ratios, leachant composition, factors controlling leachabil-

ity such as pH, redox potential, complexing capacity and physical parameters are addressed in these tests. These tests may generally take longer to perform.

2. 'Compliance' tests are used to determine whether the waste complies with specific reference values. The tests focus on key variables and leaching behaviour identified by basic characterisation tests. These tests are generally carried out within a few days at most.

3. 'On-site verification' tests are used as a rapid check to confirm that the waste is the same as that which has been subjected to the compliance test(s). These types of tests are quick quality checks, which need to be available within minutes or up to a few hours at most.

The main reason for developing a system with tests at different levels is a pragmatic one. For understanding, it is not a problem to perform a more detailed test a limited number of times. However, once the understanding of key factors is there, there is no need for the full testing any more and proper selection of the most discriminating and practical test condition can be made for regular control purposes. Although this testing philosophy was developed in relation to waste testing, it is equally applicable to construction materials[7] and sludge and soil judgement.[43]

Recently another level of test classification was identified, which has partly been addressed in Section 7.3, namely a distinction between potentially leachable, solubility controlled and dynamic test data.[44]

## 7.5.2 Characterisation of Leaching Behaviour

The types of characterisation tests that provide very useful information to assess material behaviour under a variety of exposure conditions are:

- the pH dependence leaching test, which provides a relationship of element mobilisation as a function of pH,
- a percolation leaching test, which mimics percolation behaviour, and
- a dynamic leaching test for monolithic materials, which addresses the surface-related phenomena relevant for such materials.

These three characterisation leaching tests cover more than 80% of the cases to be addressed. More specific aspects to be addressed relate to leaching influenced by reducing conditions and inorganic or organic complexants. These can best be evaluated in relation to pH as well, as these parameters are often also very pH dependent. Recent work[11,45] has addressed the mutual relationships between these characterisation tests and relationships between characterisation tests and single step extraction methods (see Section 7.5.3). Since the characterisation methods consist of more than one observation, the relationship between the data must be logical. This provides a basis to eliminate outliers or untrustworthy data. Within a material class, the major element chemistry generally does not change dramatically from one charge to another. This implies that the solubility controls are likely to be similar. This inherently leads to systematic behaviour within a materials class. The chemistry of a particular element is determined by a limited number of key parameters. Although theoretically many minerals can be formed, in practice this number is very limited for a given element. This leads to the conclusion that the generic leaching behaviour of a particular constituent should also show some generic features between different material classes with the exception of specific leachability modifying factors such as DOC, redox, *etc.* In the matrices discussed in the context of this chapter for example, the role of DOC is crucial. Another aspect that works in favour of this type of characterisation testing is that the information generated can be used more than once for different purposes and different questions, whereas the result of a single extraction test is generally short lived. Whereas results from, for instance, sequential chemical extractions are varying all the time, the results of pH dependence tests are much more stable and repeatable. This implies that it will be important to make this information generally available, for instance in the form of a database.

In Figure 6 the relationship between the pH dependence leaching tests and the percolation leaching test is illustrated for Cr and Cd in the case of heavily sewage sludge-amended soil (CSO2). For comparison the total composition is also given by the broken line in the left hand graphs. As indicated before, the release as function of L/S can be related to a time scale through the infiltration rate, the height and density of the material. The dotted line in the right hand graphs indicates solubility control. When the data points start to deviate strongly from this slope 1 (*e.g.* Cd) and start to approach a slope 0, then depletion of a soluble form is observed. As the chemical environment may be different from one case to another, the behaviour may vary between materials and between constituents.

## 7.5.3 Comparison of Test Methods

The pH dependence test has been marked as a method that allows mutual comparison of different test methods.[10] This aspect of the method is illustrated in Figure 7 and Figure 8 for a random selection of elements from contaminated harbour sediment (SED3) and from a heavily sewage sludge-amended soil (CSO2). The agreement between the different tests – $CaCl_2$ extraction,[46] EN 12457,[47] $NaNO_3$ extraction,[48] acetic acid extraction[49] and sequential chemical extraction[50] – carried out on the same materials is highly encouraging. Obviously, the EDTA extraction[51] does not fit the pH pattern as EDTA complexation is almost independent of pH. In fact, very similar results to EDTA extraction can be obtained at low pH, so if the aim is to assess an availability level, extraction at low pH would suffice. The latter option is even better as EDTA is not equally effective for all constituents of interest and an extraction at low pH is more effective for a wide range of elements. The results from the sequential chemical extraction are calculated as a cumulative leached quantity at the pH corresponding to the subsequent more aggressive extraction step. Particularly, in the case of the sediment where the pH dependence test was carried out over the full pH range covered by all other test methods, the agreement is very good. For comparison purposes the total composition is included as a horizontal broken line. The curve obtained with the pH dependence test can thus be seen as a

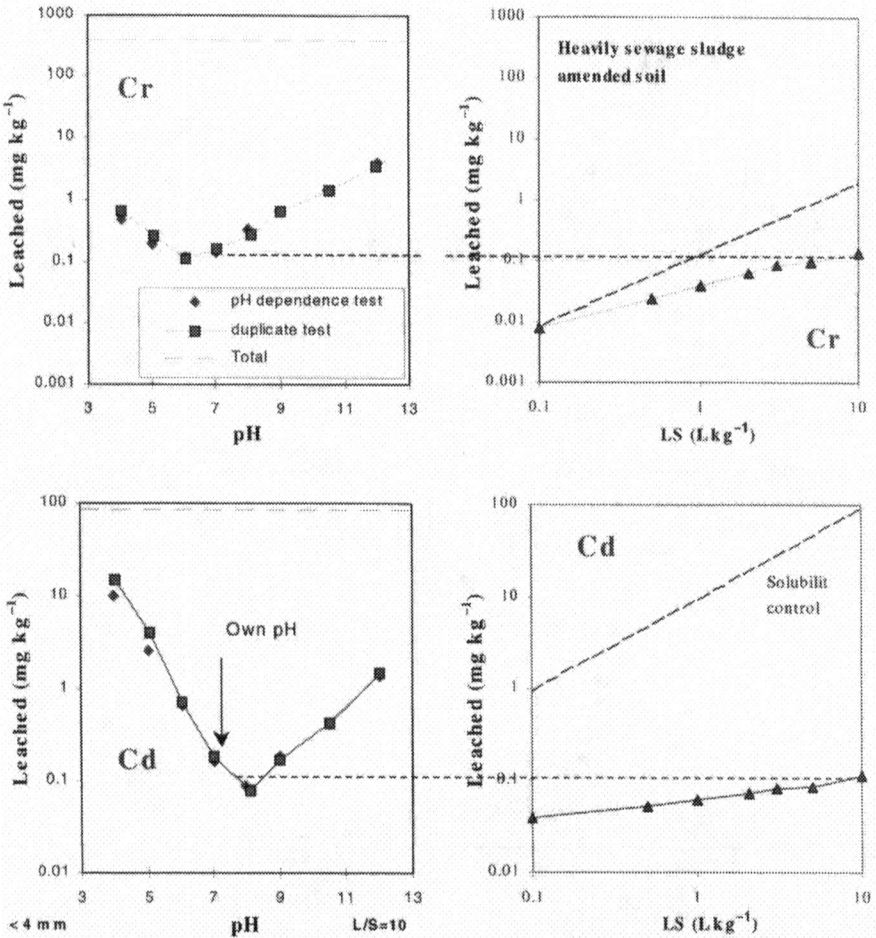

**Figure 6** *Relationship between percolation test and pH dependence test for contaminated soil (CSO2) resulting from excessive sewage sludge amendment (SMT4-CT96-2066).[11] Dotted line in the right hand graphs indicates solubility control. Broken line in the left hand graphs indicates total composition. This information is important to assess long term leaching behaviour and to identify release-controlling mechanisms. The pH dependence test provides information on changes in conditions, whereas the release as a function of L/S covers the time dependent aspect*

material characteristic. This applies to metals, oxyanions, major and minor elements and DOC. Here, only a limited number of parameters are shown, but the same observation extends a wide range of other constituents.[11]

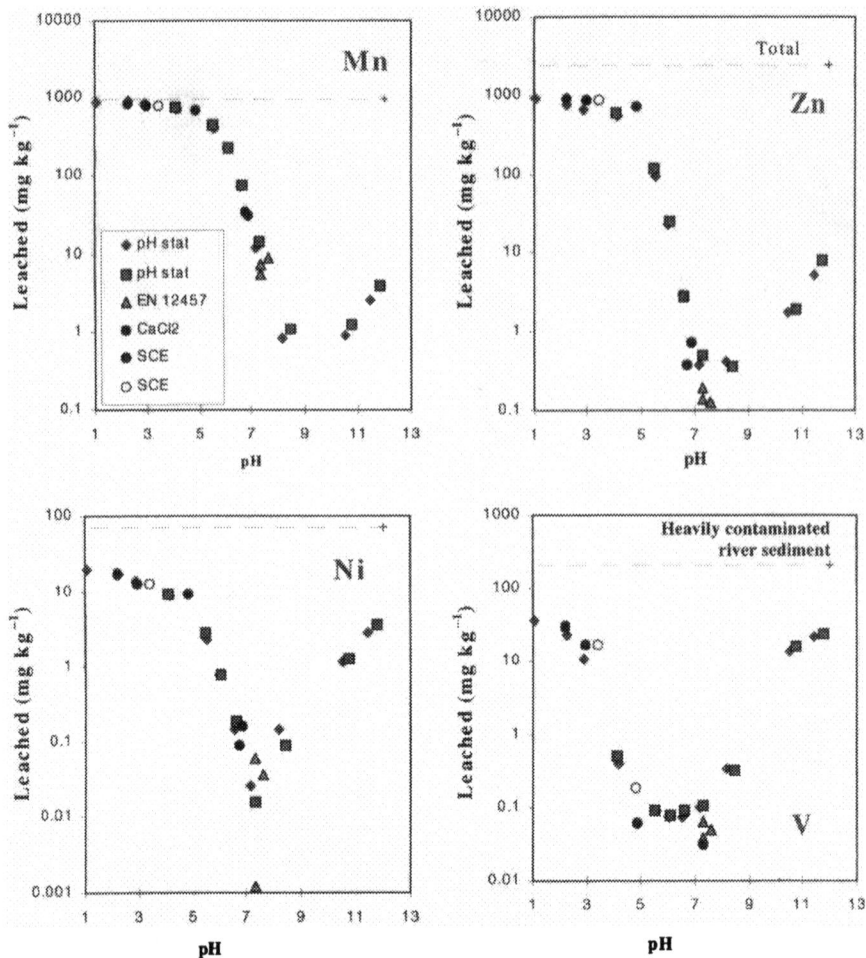

**Figure 7** *Contaminated river sediment (SMT4-CT96-2066).[11] Comparison between a range of leaching tests procedures: pH dependence test, EN 12457-2, CaCl₂ extraction and sequential chemical extraction. Different tests can be placed in perspective by plotting them in relation to the pH dependence test. Even SCE matches well when plotted as a cumulative leaching series*

## 7.5.4 Quality Control System

Once a sufficient level of understanding of the key controlling factors are known, single extractions or a limited number of extractions will suffice for quality control purposes. Single extraction test results on their own are meaningless. By

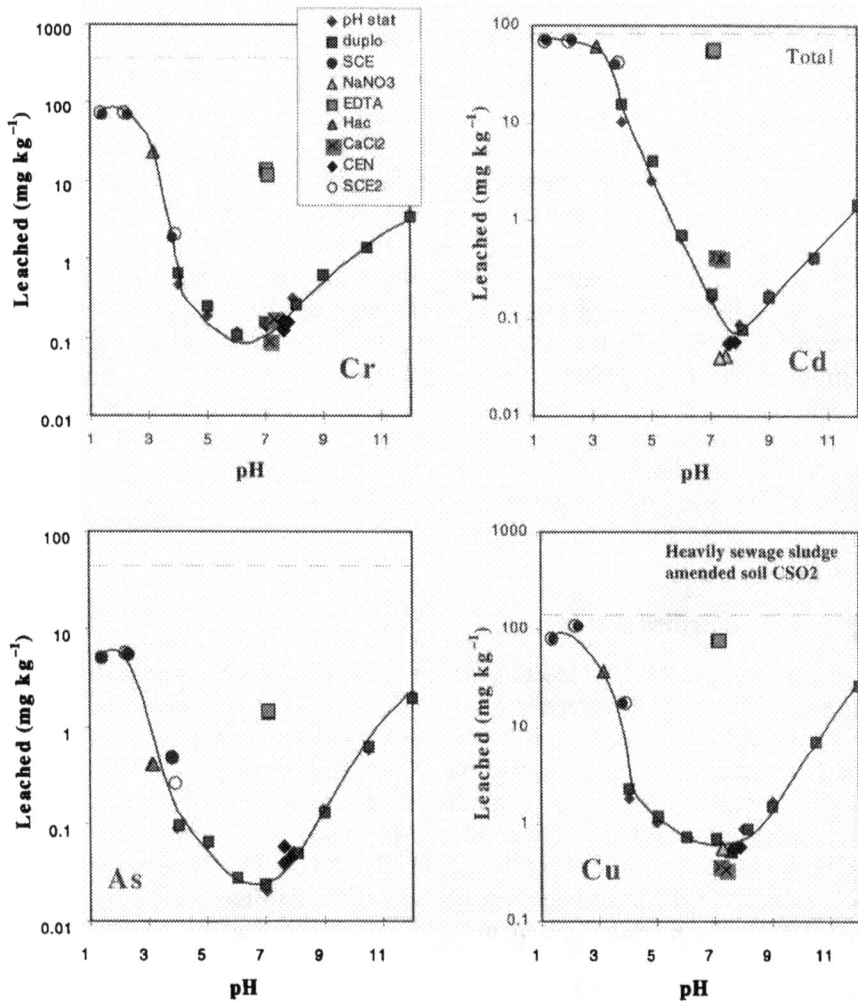

**Figure 8** *Contaminated soil heavily amended with sewage sludge (SMT4-CT96-2066).*[11] *Comparison between a range of leaching tests procedures: pH dependence test, EN 12457-2, CalCl$_2$ extraction, NaNO$_3$ extraction, acetic acid extraction, EDTA extraction and sequential chemical extraction. Different tests can be placed in perspective by plotting them in relation to the pH dependence test*

always presenting compliance or quality control test data in conjunction with the associated characterisation information conclusions can be drawn on compatibility of the data of the batch under investigation with the previously determined leaching character. Deviations from the normal leaching character can be the result of an analytical error or a true deviation in chemical speciation. In the latter case, further verification may be needed to ensure what has caused the deviation.

## 7.5.5   Prediction of Release under Different Exposure Conditions

The pH dependence leaching tests in conjunction with the percolation test can provide insight into long-term changes in leaching behaviour. For instance, materials with a pH deviating from that of most natural environments will start to change with time at a rate depending on the buffer capacity of the material. In Figure 9 different exposure conditions are reflected by the different release levels in the percolation test corresponding to the release levels in the pH dependence leaching test. This information is applicable in situations where the release is largely solubility controlled and local equilibrium can be expected. In the case represented in Figure 9 the initial alkaline conditions of the material will be neutralised under the influence of contact with atmospheric $CO_2$ and $CO_2$ originating from biological degradation. For a material with a layer thickness of 1 metre and an acid neutralisation capacity of 0.2 mol kg$^{-1}$, it may take less than 10 years for neutralised conditions to be established. Given the large difference in leaching behaviour of several elements, it is important to know when judging materials for long term environmental impact.

## 7.5.6   Standardisation Issues

To ensure comparability of leaching test results for the same material, standard tests methods need to be agreed upon. In CEN and ISO, standards are developed by many technical committees and working groups on a wide range of subjects. The CEN and ISO TCs are organised vertically. This implies that relatively little interaction may occur between committees working on similar subjects. In the last decade, this aspect has been criticised, which has led to some actions to try and bridge existing gaps between TCs. Recently, the EU Environment Directorate has emphasised the need for more horizontal standardisation as a regulatory need. Technical committees active in the development of leaching test methods

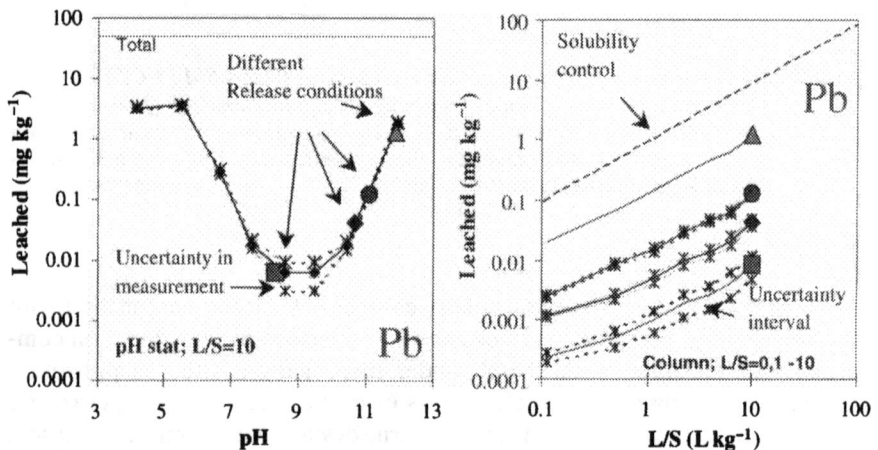

**Figure 9**   *Relationships between pH and L/S (time dependent release)*

have been summarised a few years ago.[10] In the field of sludge, biowaste and soil, the relevant Committees today are: ISO/TC 190 Soil – Sub Committee 7 Soil and Site Assessment – Working Group 6 Leaching. In this group three methods are being developed; a percolation test for soil and two batch leaching methods at low and higher liquid to solid (L/S) ratios. CEN/TC 308 Characterisation of Sludge, which has adopted the batch leaching tests developed by CEN/TC 292 Characterisation of Waste, where leaching tests are developed at the three hierarchical levels mentioned before. In Working Group 2, the compliance leaching test EN 12457 has been developed in four parts (different L/S ratios and two particle size ranges) and recently been validated with more than 50 participating European laboratories. In Working Group 6, basic characterisation methods are developed. A percolation test for granular waste is about to become a formal European standard.[27] The same applies for a pH dependence leaching test as a series of batch extractions with different initial acid/base additions to generate a full pH dependence leaching curve.[28] A dynamic leaching test for monolithic materials is still in preparation.[52] Characterisation without interpretation and modelling is not very useful. So an important aspect of characterisation test data is their use as input to modelling.

# 7.6 Modelling of Chemical Speciation and Modelling Release

Modelling can be carried out for different purposes. It can be important to identify the solubility controlling phases by means of equilibrium based models such as the geochemical speciation codes MINTEQA2 from Allison $et\ al.$[53] and ECOSAT.[54] For evaluation of release in a given scenario, modelling the dynamic release from a material under specific exposure conditions will be relevant. In this case, dedicated models focused on a combination of chemical reaction and transport are relevant.[55]

Both types of modelling are important to reach the understanding of processes controlling release in a given scenario. First geochemical equilibrium modelling will be addressed and then an example of chemical reaction/transport modelling will be addressed.

## 7.6.1 Geochemical Modelling

For two contaminated soils,[45] the ion activities were calculated using the Davis equation based on the pH dependence leaching tests data as input to the model. The modelling approach consisted of calculating the saturation indices (SI) of all available minerals in MINTEQA2.[53] Potentially solubility-controlling minerals were selected on the basis of the calculated saturation indices ($-1 < SI < 1$). The minerals that gave the best prediction of the experimental pH dependent solubility curve were selected. Lines have been drawn assuming equilibrium ($c_{eq}$) with minerals by calculating the quotient of the experimental concentration ($c_{det}$) of an element and the calculated saturation index as $10^{SI}$: $c_{eq} = c_{det}/10^{SI} = c_{det}$ (IAP/

$K)^{-1}$ (IAP is the calculated ion activity product on the basis of the experimental data; $K$ is the stability constant).

This modelling approach is a slight simplification of the method used by Meima and Comans[56] because in the calculations here it is assumed that the speciation does not change significantly for situations where the saturation index is in the range of $-1 < SI < 1$ and $SI = 0$, for which the equilibrium of a mineral is drawn as a line. In the method used by Meima and Comans[56] the speciation in solution was calculated separately for each mineral by assuming for each mineral an infinite amount of the mineral. Both methods give approximately the same results when the initially calculated saturation indexes are approximately $-1 < SI < 1$. In Figure 10 the results of the modelling are given in comparison with the original data from the pH dependence leaching test.

The match between the model prediction and the measured data is within the specified SI range for significant portions of the pH domain. In addition the shape of the model curve largely matches the measurement, which is also indicative of potential solubility control by the $BaSO_4.BaCrO_4$ solid solution. This phase has been found to be potentially solubility-controlling in many different matrices.[7,57] When the solubility-controlling phase can be identified, more realistic predictions of long term behaviour of the material can be made.

### 7.6.2  Modelling of Metal Organic Matter Interactions

For organic-rich matrices the interaction with particulate and dissolved organic matter is crucial to describe metal leachability. Here, results of modelling metal–organic matter interactions for sewage sludge, compost and contaminated soil is presented as an example of the current possibilities to predict release from organic matter-rich matrices. Six materials have been investigated in the 'Technical Work Harmonisation of Leaching/Extraction Tests'[11] in which the metal

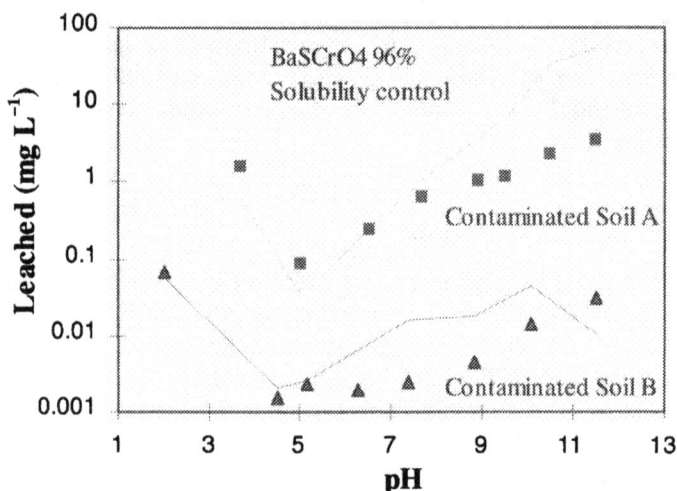

**Figure 10**   *Modelling results for Cr leaching from two contaminated soils*

concentrations and DOC concentration has been determined as a function of pH (according to CEN/TC 292-WG6).

The NICCA–Donnan model has been developed by Koopal *et al.*[58] and Kinniburgh *et al.*[59,39] to be able to model the metal binding on natural organic matter. The NICCA–Donnan model is a semi-empirical model for the calculation of chemical equilibria on natural organic matter, which can be used in standard models to calculate chemical equilibria, such as ECOSAT.[54] In the model, the assumption is made that the chemical affinity for protons and metals is caused by sites such as carboxylic and phenolic groups, with certain site densities (Q) that have an affinity for protons and metals that is not discreet but is distributed around certain peak values. It follows that each kind of ion has certain distributions of chemical affinity around peaks (log $K$) where the distribution of chemical affinity is given by a parameter $p$. Besides a heterogeneity for each ion there is a heterogeneity that is non-specific and general ($n$). The electrostatic part of the affinity of ions for organic matter is regulated in the NICCA–Donnan model by a Donnan model for which it is assumed that the charge of the organic matter is fully compensated by counter charge inside a certain Donnan volume. The Donnan volume is a function of the charge or pH, and salt strength. Seeing that the Donnan volume hardly changes as a function of the pH above a salt strength of 10 mM, the Donnan volume is described by an empirical relation as a function of the salt strength.[60,40]

The pH dependent leaching of heavy metals from several materials has been modelled using the NICCA–Donnan model, which is incorporated in ECOSAT, a computer model to calculate chemical equilibria. The modelling of the pH dependent solubility by assuming mineral equilibria, as shown earlier, does not give a good description of the leaching of the heavy metals, especially at neutral to high pH values. The NICCA (non-ideal consistent competitive adsorption) model parameters for an extracted and purified peat humic acid were used as given by Kinniburgh *et al.*[39] except for the site density, which was taken from Benedetti *et al.*[40] when they model field data. The site density has been optimised for each material to give a good fit of the data as this factor has been found to vary between different materials. The factor size of the organic molecules has not been taken into account as has been done earlier to model the acid–base behaviour by de Wit *et al.*[61] and Avena *et al.*[62] The effect of the factor size of DOC for metal binding is unknown.

The following assumptions were made for the model calculations:

1. Model parameters:
   a. The NICCA model parameters as determined for Ca, Al, Pb, Cd, Zn and Cu by Kinniburgh *et al.*[39] have been used except for the site density.
   b. It has been assumed that the site density for DOC and organic matter for each material is the same and is equal to a certain fraction of the site density used for PPHA by Kinniburgh *et al.*[39] Also the distribution over the two types of sites (the carboxylic and phenolic type has been kept constant). The adjustment of only the site density has been

used successfully by Benedetti *et al.*[40] to describe the metal binding on a lake fulvic acid and soil organic matter. It has been assumed that clay minerals do not have a significant influence on the metal speciation in the materials used. For the calculation the speciation in solution has been taken into account (for example metal–phosphate species).

2.  Input parameters:
    a.  An equilibrium with gibbsite has been assumed as this mimics the experimentally found pH dependent Al concentration and Al included in the NICCA–Donnan model and is able to compete with Ca, Pb, Zn, Cd, and Cu.
    b.  Speciation in solution: the determined amounts of Na, Mg, Ca, P, S and, if determined, TIC have been included in the calculations.
    c.  The concentration of dissolved organic carbon (DOC) has been taken into account in the calculation.
    d.  A certain amount of organic matter has been assumed, including the amount of DOC that is precipitated at low pH values (so the total amount of carbon is constant for each material).
    e.  The total concentrations of Cu, Pb, Zn and Cu have been chosen for each material on the basis of the highest concentration found in solution, or slightly more if it was necessary to get a better description of the data.

As the research is still in a pilot stage, the optimisation of the model parameters has been done without a mathematical routine to reduce the difference between the experimental and calculated results. Today, a mathematical routine is available (FIT)[63] that enables us to optimise all the model parameters for one set of experimental data for one metal.

The results show that the site density is a sensitive parameter for the pH dependent data for which the modelling has been performed. Different site densities have been assessed for the different materials. The site density of PPHA has been determined by optimisation of the acid–base behaviour by Kinniburgh *et al.*[39] and was found to be 6.64 mol kg$^{-1}$, while the optimised site densities for CSO2 and IPF are equal to that. In case of CW1 and CW5, values of respectively $\frac{1}{3}$ and $\frac{1}{2}$ were used. These values are in the range to be expected.

The model calculations for the four materials with widely different organic matter contents show a good description of the Pb, Zn, Cu and Cd concentrations in solution as illustrated in Figure 11. For IPF the description of the Cu, Pb and Zn concentration is too high which means that the bonding of Pb and Zn to the DOC is overestimated or the bonding by the soil is underestimated. Without measurements of the bonding characteristics of the separate phases – DOC and the soils without DOC – the calculations can not be tested. The good description of the metal concentrations in solution with increasing pH suggests that binding to DOC is, as assumed, the process responsible for the increase of the solubility as a function of pH at pH values higher than 7.

Modelling the metal binding in soils, sediments and waste materials is still in a

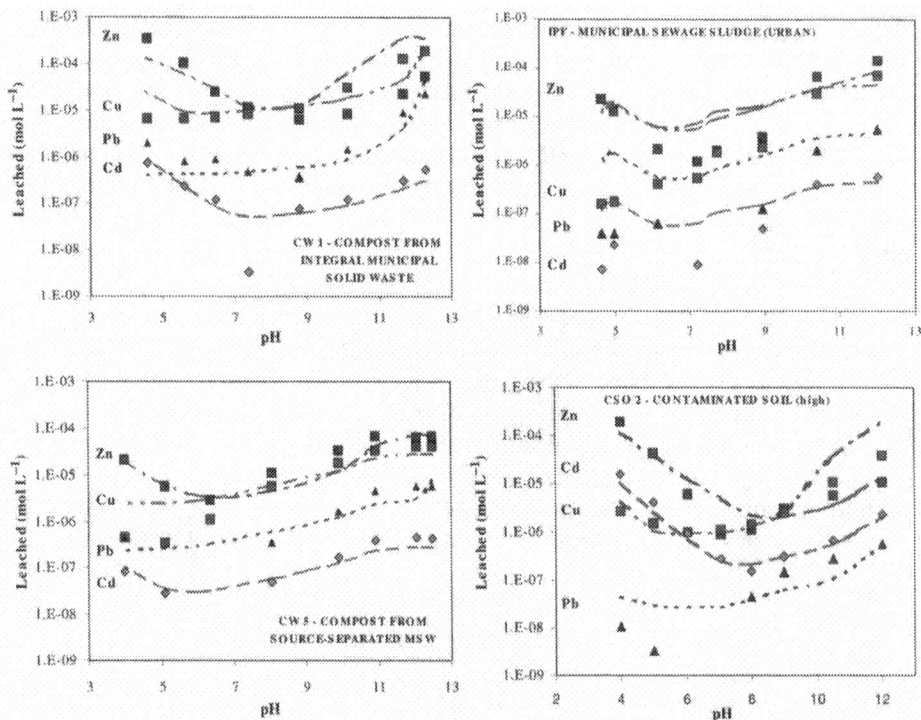

**Figure 11**   *Modelling of metal–organic matter interaction (both particulate and dissolved) in contaminated soil using NICCA–Donnan model implemented in ECOSAT illustrating that metals release is dominated by organic matter interaction and DOC complexation[11]*

pilot stage but the results show a good perspective for future research and characterisation. The description of the pH-dependent metal concentrations with a minimum of adjustable parameters indicates that the NICCA–Donnan model offers a very good perspective to model metal binding to soils and waste materials with natural organic matter. It is of particular interest as it is able to model the complex distribution of metal over dissolved and particulate material. The results suggest that with one model the data for different materials can be modelled by using one unique set of parameter values.

## 7.7   Comparison of Sewage Sludge-amended Soils and Sewage Sludges

From an environmental impact point of view, it is interesting to be able to relate the quality of sewage sludge to its impact when mixed in a certain proportion with soil. This impact is twofold. It relates to uptake of possible contaminants from sludge by plants, which may enter the food chain by this route. The other option is the release of contaminants from sludge-amended soil, which may affect the underlying soil system and ultimately may affect groundwater. The leaching test results, as obtained by the pH dependence leaching tests and the associated modelling, provide a means of quantifying the relationship between sludge composition and the sludge-amended soil. The key factor, as we have seen in the previous chapter, is the nature of the organic matter and the interaction of metals with particulate and dissolved organic matter. A first comparison given here is the comparison of two sludge amended soils (CSO2 – a heavily sewage sludge-amended soil[11] and CRM 483[64]). In Figures 12 and 13 the pH dependence test data for CSO2 and CRM 483 (see Chapter 3) are given for respectively Cd, Cr, Cu, DOC and Ni, Pb, Zn and Ca. The leaching curves for the two sludge-amended soils is very similar in shape for all elements presented here. For DOC, Cr, Cu, Ni and Pb, the shape of the curves is very similar. However, the leached quantities in the pH range > pH 5 are different for these elements. They are apparently affected by the significantly increased DOC level in the CRM 483 soil. The affinity of the individual elements for the respective sorption sites on the more abundant organic matter in one *versus* the other material determines which elements will show an increase and for which it is less prominent. For instance,

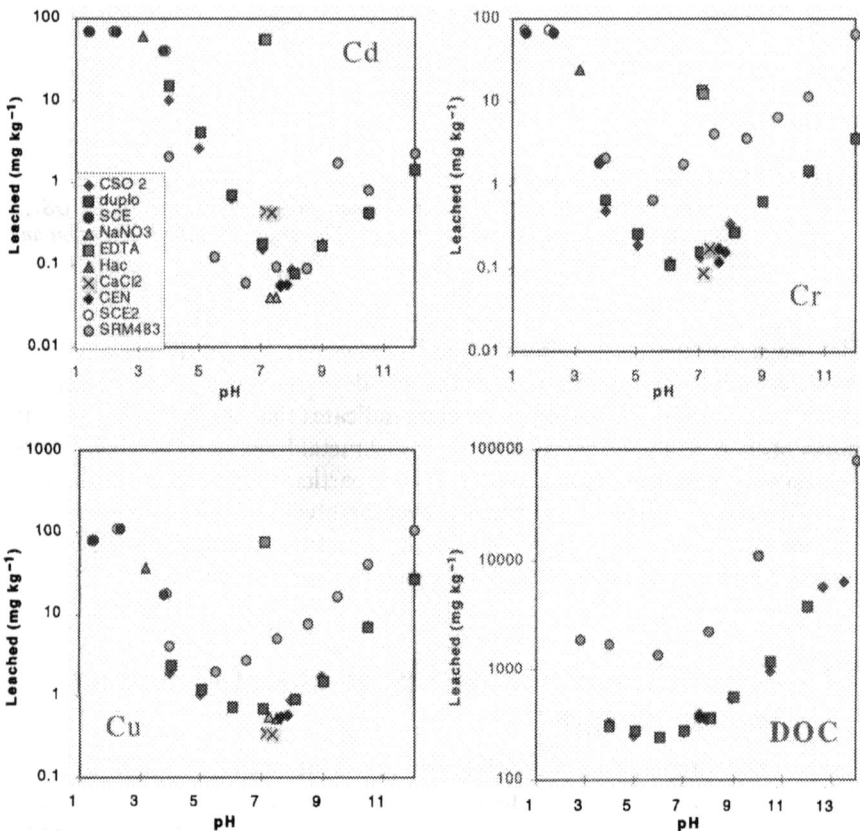

**Figure 12**   *Comparison of the leachability of Cd, Cr, Cu and DOC from heavily sewage sludge-amended soils CSO2 and SRM 483*

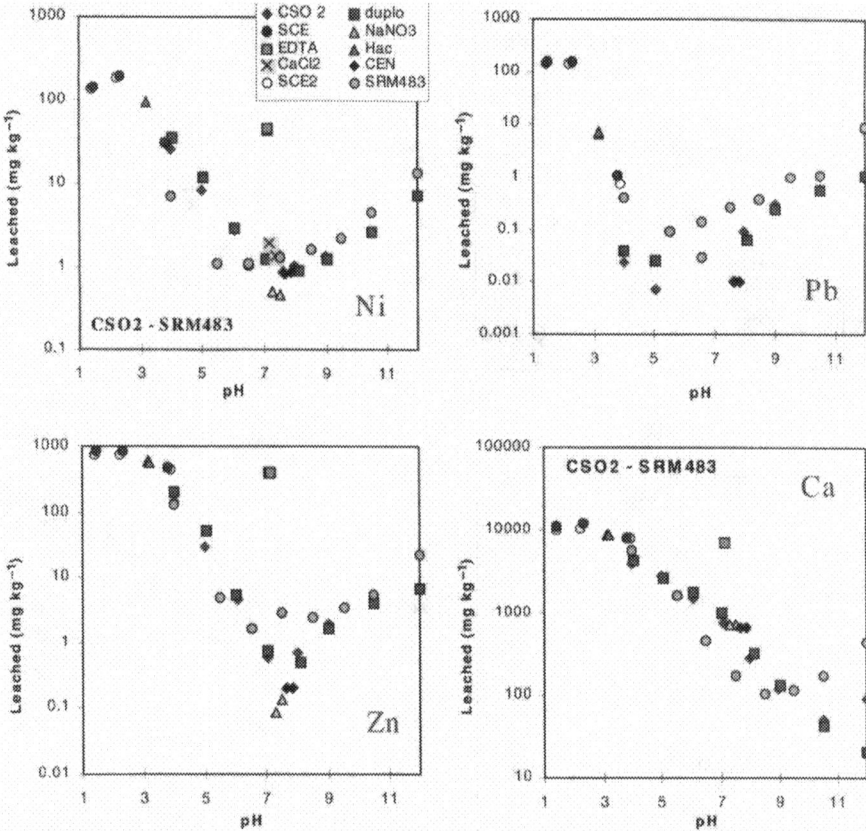

**Figure 13**  *Comparison of the leachability of Ni, Pb, Zn and Ca from heavily sewage sludge-amended soils CSO2 and SRM 483*

Zn has a much lower affinity than Cu, which is consistent with the observed responses to an increased DOC level.

The comparison of sewage sludge-amended soil with the leachability of sewage sludge as shown in Figure 14 illustrates similarities between urban sludge leaching behaviour and heavily sludge-amended soil behaviour. For rural (more organic) sludge, the leachability is markedly different. The rural sludge is characterised by a significantly higher degradable and dissolved organic matter content. At the same time the metal content is significantly lower in the rural sludge. It should be noted that at low pH (pH < 4) the leachability of Zn is definitely lower in comparison with the urban sludge and in the case of Cu the leachability is still decreasing in this pH range. This is contrary to what one would expect for these metals. The explanation seems to be that there is an abundance of particulate organic matter in the rural sludge which is very effective in scavenging metals from solution, even at low pH.

Data for urban French and urban Dutch sludges match very well. Zn and Cu

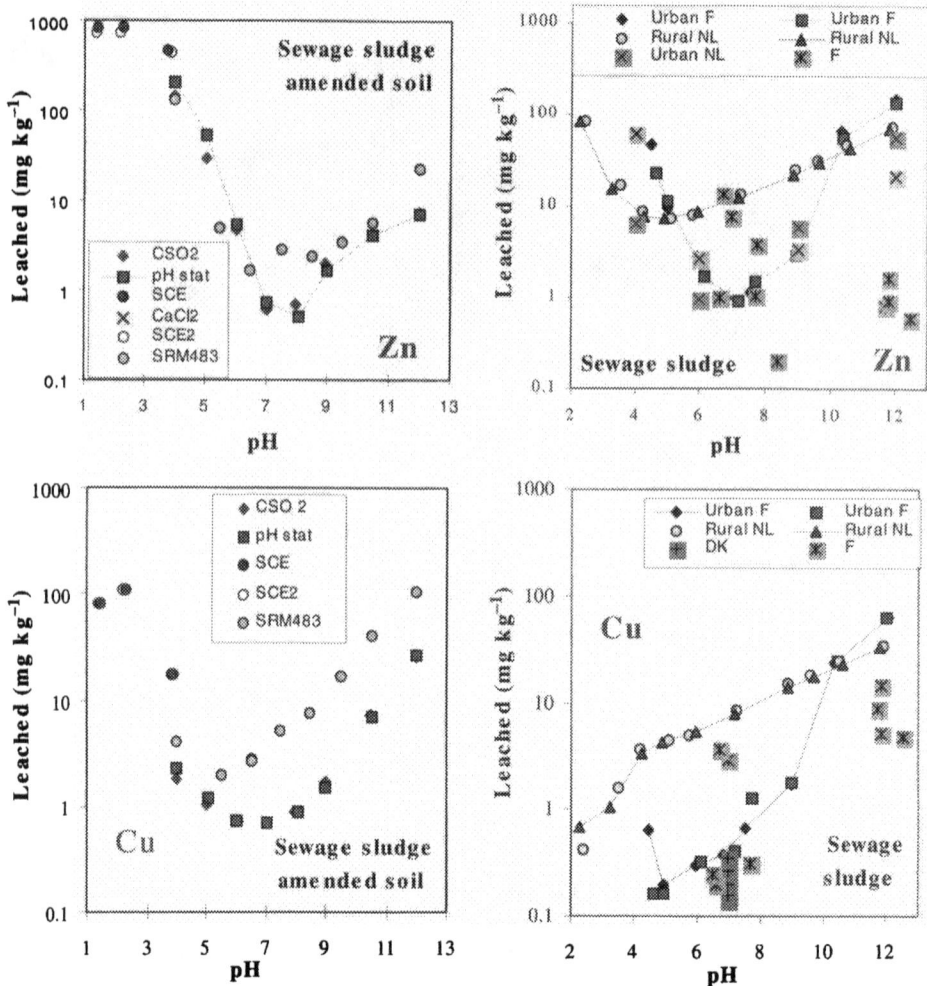

**Figure 14**  *Comparison of leaching of Zn and Cu from sewage sludge-amended soil and leaching from raw urban and rural sewage sludge*

leaching of individual sludge samples as obtained by single step extraction methods matches with data for urban and rural sludge.

Based on the modelling[11,55] and the results for the various materials containing organic matter, it follows that the mobilisation of metals at low pH (pH < 5) is largely governed by the sorption capacity on particulate organic matter. As soon as the metal loading increases, the metals show an increase in the order of least affinity for the organic matter sorption sites. This implies that, for instance, Zn is released first and Cu last. At the intermediate pH level (pH 5–7) mobilisation is already affected by DOC. In the high pH range (pH > 7) metal mobilisation is controlled by association with DOC. Increased DOC levels in materials are reflected in the degree of metal mobilisation. The metal mobilisation as a function of pH proceeds parallel to the increase in DOC. The metal loading and the relative proportion of the metals dictates the extent of DOC mobilisation for a specific element. This implies that there is no direct proportionality between individual metal mobilisation and DOC level changes. In Figure 15 the leaching behaviour of a range of soils, contaminated soils and sewage sludge-amended soils is compared, which illustrates the systematic behaviour as described above. Obviously, at low concentration levels the scatter in leaching test data is larger. The parallel curves for Cu at pH > 7 at different DOC loadings are noticeable.

## 7.8 Judgement of Test Results and Regulatory Developments

The use of leaching tests for regulatory purposes has traditionally been limited to single extraction test data. As a consequence, a lot of studies to evaluate material quality have been limited to these simple tests, which provide no information on key controlling factors. Characterisation tests as standardised now in CEN/TC 292 provide insight into such factors and offer a better potential for treatment of material, which would otherwise not be obvious. A concise test has been developed,[68,69] in which key aspects of leaching (pH and L/S dependence) are covered in a just a few extractions. The fear that characterisation is too costly is misplaced

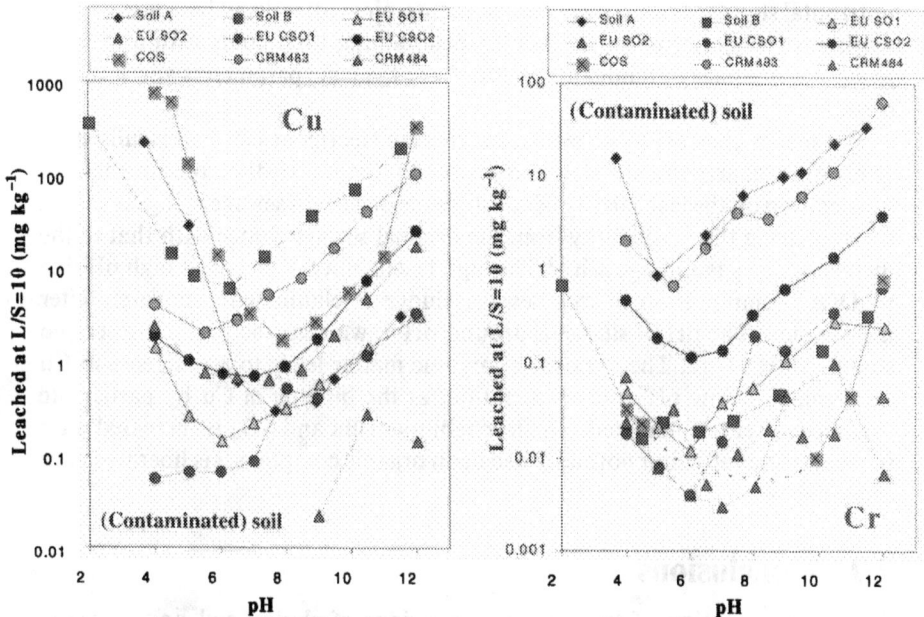

**Figure 15**  *Comparison of contaminated soil and sludge-amended soil leaching data illustrating the consistent behaviour. Soils A and B are Austrian contaminated soils,[45] EU SO1, SO2 are EURO soils,[65,11] CSO1 and CSO2 are heavily sludge-amended soils,[11] COS is an industrially contaminated soil used in the CEN/CT 292 validation study,[66] CRM 483[64] and CRM 484[67] are sewage sludge-amended soils (see further details in Chapter 3 for these latter two soils)*

as characterisation is not carried out with the same frequency as quick screening tests. In addition, the characterisation tests provide a better basis for the decisions that need to be made and provide a key to previously unexplained differences in the single-step tests. A cost effective solution would be to apply characterisation testing to a material class and use that information as basis of reference for single-step compliance or screening tests.

The development of environmental criteria for biomaterials relates to the utilisation of sludge and compost as fertilisers, as biofuel and, in case these materials are too contaminated for such beneficial application, disposal in landfills. In each of these scenarios regulations are in force or in development.[31,70,71] A key aspect of efficiency is that materials tested for one application do not have to be tested extensively to verify their acceptability for other uses or disposal. This harmonisation of methods across different fields is gaining interest as it is a source of dispute and a major source of additional cost. The relevant European directives for the various applications of soil and biowaste are the EU Sludge Directive[70] and the Working Document on Biowaste[71] as basis for the new Biowaste Directive, the Council Directive 1999/31/EC on Landfill of Waste[31] and a Biofuel directive (in preparation). The Sludge and Biowaste directives focus mainly on total composition, whereas the Landfill Directive focuses mainly on leaching. Recently, initiatives have been taken to develop horizontal standards for sewage sludge, (contaminated) soil and biowaste.[72] This work covers harmonised methods for determining total composition of inorganic and organic contaminants, methods for hygienic parameters and leaching methods.

The major changes as a result of a specific treatment can be visualised by applying a characterisation leaching test. In Figure 16 the consequences of incineration on the leachability of constituents present in sewage sludge is shown by comparing the leachability from the original sewage sludge with that of the incinerated sewage sludge ash.[11] The high leachability of metals at high pH due to DOC complexation in raw sewage sludge is eliminated completely after incineration. The pH of sludge is around pH 7, whereas the pH of incinerated sludge is pH 11–12. The removal of organic matter leads to an increase in Cu leachability at low pH after incineration, as the binding of Cu by particulate organic matter is eliminated. The leachability of Ba and $SO_4$ is increased after incineration as all sulfur normally tied up in organic compounds is liberated after incineration.

## 7.9  Conclusions

The characterisation of the leaching behaviour of sludge, soil and biowaste indicates that the leaching behaviour of materials with a relatively high organic matter content is quite systematic. The pH dependence leaching test proves to be one of the most useful methods to characterise the material behaviour under a variety of exposure and treatment conditions. It provides a basis of comparison for almost any other existing leaching test with the exception of EDTA extraction. The latter has a serious limitation when it is intended as a means of

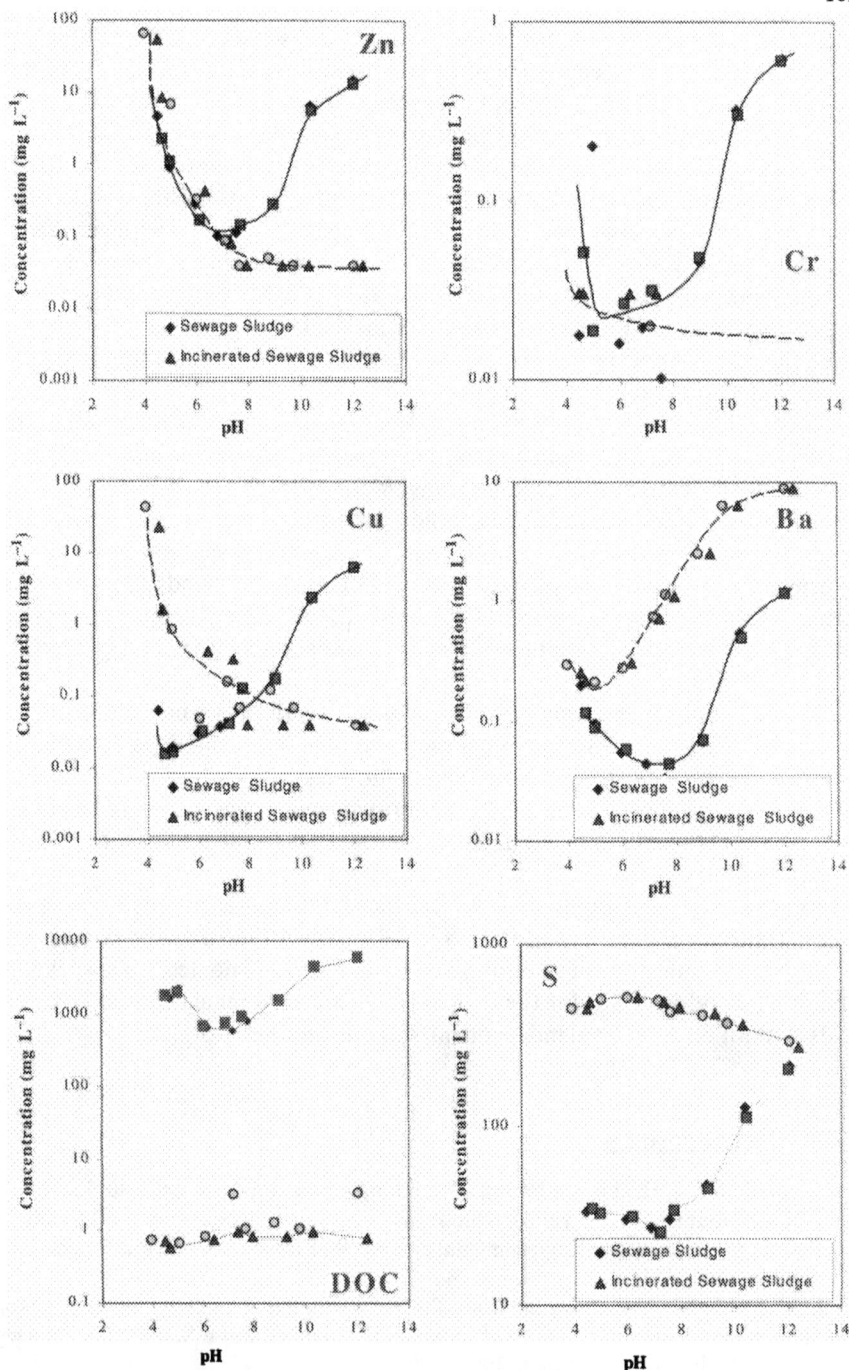

**Figure 16** *Comparison of leaching behaviour of sewage sludge and incinerated sewage sludge. pH Sewage sludge: 7.3, pH incinerated sewage sludge: 7.9. Indication of use of pH dependence test to evaluate consequences of treatment methods.* ▲ *and* ●*, and* ■ *and* ◆ *are duplicate test results*

assessing availability for plant uptake, as it does not have the same complexing behaviour towards all elements of interest. In addition, the pH dependence test provides a means of deriving chemical speciation aspects, either by experience (previous observations) or by modelling using the test data as input to geochemical speciation codes. Unlike the sequential chemical extraction method which claims speciation and only provides qualitative interpretations, the pH dependence test can provide true speciation information in combination with geochemical codes (MINTEQA2, ECOSAT) able to quantify mineral solubility and specific codes to quantify sorption on metal oxides (Fe, Mn oxides) and element – organic matter interaction (NICCA–Donnan). In the organic matter-rich matrices, the role of DOC is one of the most dominant factors controlling release of metals, but also organic contaminants.

The level of agreement between metal leaching behaviour as obtained in the pH dependence test and modelling of metal–particulate matter and metal–dissolved organic matter interactions using model parameters, which were not obtained for the materials studied, holds promise for the future, when such material specific parameters (*e.g.* site density) have been established. When the agreement can be further improved, then predictions beforehand are possible on certain levels of organic matter and DOC. This allows setting targets for required performance of treatment methods such as composting, biological degradation, *etc.*

The prediction of the long-term behaviour of materials containing organic matter has long been considered impossible. These recent developments in combination with dynamic testing (percolation test) and associated modelling provide the basis for more detailed long-term assessment of impact from sludge use, compost application, contaminated sites and new applications and reuse of (contaminated) soil.

From an efficiency point of view, the need for consistent methods to be referenced in regulation horizontal standardisation will be a major issue in the next decade. A multitude of tests for different matrices will not lead to a workable situation for industry nor for the regulatory agencies as regulations in adjacent fields can not be developed independently of one another.

# 7.10   References

1. Thematic Network Harmonisation of Leaching/Extraction tests 1995–2001.
2. J.J.J.M. Goumans, H.A. van der Sloot, Th.G. Aalbers, (eds.), *Waste Materials in Construction*, Studies in Environmental Science 48, Elsevier Science Publishers, Amsterdam, 1991, 672.
3. J.J.J.M. Goumans, H.A. van der Sloot, Th.G. Aalbers, (eds.), *Environmental Aspects of Construction with Waste Materials*, Studies in Environmental Science 60, Elsevier Science Publishers, Amsterdam, 1993, 988.
4. J.J.J.M. Goumans, G.J. Senden, H.A. van der Sloot (eds.), *Waste Materials In Construction – Putting Theory into Practice*, Studies in Environmental Science 71, Elsevier Science Publishers, Amsterdam, 1997, 886.
5. H.A. van der Sloot, O. Hjelmar, J. Mehu and N. Blakey, in *Proceedings Sardinia 99*,

*Seventh International Landfill Symposium*, S. Margharita di Pula (ed.), Cagliari, Italy, 1999, 3–10.

6. IAWG (International Ash Working Group; A.J. Chandler, T.T. Eighmy, J. Hartlen, O. Hjelmar, D.S. Kosson, S.E. Sawell, H.A. van der Sloot, J. Vehlow), *Municipal Solid Waste Incinerator Residues*, Studies in Environmental Science 67, Elsevier Science, Amsterdam, 1997, 974.

7. H.A. van der Sloot, *Cement & Concrete Res.*, 2000, **30**, 1079.

8. H.A. van der Sloot, R.N.J. Comans and O. Hjelmar, *Sci. Total Environ.*, 1996, **178**, 111.

9. P.M. Esser, H.A. van der Sloot, and W.L.D. Suitela, *Harmonisation of Leaching Tests: Leaching Behaviour of Wood*, Heron, 2000, **46**(4), 15–19.

10. H.A. van der Sloot, L. Heasman and Ph. Quevauviller (eds.), *Harmonization of Leaching/Extraction Tests*, Studies in Environmental Science, Volume 70, Elsevier Science, Amsterdam, 1997, 292.

11. Technical Work in Support of the Network Harmonisation of Leaching/Extraction Tests, Contract SMT4-CT96-2066, European Commission, Brussels, Belgium.

12. H.A. van der Sloot, *Harmonisation of Leaching/Extraction Tests and Leaching of Organic Contaminants*, contribution to TRA-EFCT meeting 'Clean Technologies', Brussels, November 1998, ECN-RX-98-068, 1998.

13. *Leaching of Organic Contaminants*, final report, Contract SMT4-CT97-2160, European Commission, Brussels, Belgium.

14. Chapter 6 of this book.

15. H.A. van der Sloot, *Waste Manag.*, 1996, **16**, 65.

16. D.S. Kosson, H.A. van der Sloot and T.T. Eighmy, *J. Hazard. Mat.*, 1996, **47**, 43.

17. P. Moscowitz, R. Barna, F. Sanchez, H.R. Bae and J. Mehu, 'Models for Leaching of Porous Materials', in *Waste Materials In Construction – Putting Theory into Practice*, J.J.J.M. Goumans, G.J. Senden and H.A. van der Sloot, (eds.), Studies in Environmental Science 71, Elsevier Science Publishers, Amsterdam, 1997, 491–500.

18. O. Hjelmar, H.A. van der Sloot, D. Guyonnet, R.P.J.J. Rietra, A. Brun and D. Hall, in *8th Waste Management and Landfill Symposium*, October 2001, Volume III, 721–771.

19. T. Taylor Eighmy, D. Crimi, I.E. Whitehead, X. Zhang and D.L. Cress, 'The Influence of Monolith Physical Properties on Diffusional Leaching Behaviour of Asphaltic Pavements Constructed with MSW Combustion Bottom Ash', in *Waste Materials In Construction – Putting Theory into Practice*, J.J.J.M. Goumans, G.J. Senden and H.A. van der Sloot, (eds.), Studies in Environmental Science 71, Elsevier Science Publishers, Amsterdam, 1997, 125–148.

20. Building Materials Decree, Staatsblad van het Koninkrijk der Nederlanden, 567, 1995.

21. EPA Toxicity Test Procedure (EP-tox), Appendix II, Federal Register, Vol 45(98), 1980, 33127–33128. Government Printing Office, Washington DC later changed to Toxicity Characteristic Leaching Procedure (TCLP). Federal Register Vol 51, No 114, Friday, June 13, 1986, 21685–21693 (proposed rules). Federal Register, Vol 261, March 29, 1990 (final version).

22. DIN 38414 S4: German standard procedure for water, wastewater and sediment testing – group S (sludge and sediment); determination of leachability (S4). Institüt für Normung, Berlin, 1984.

23. Déchets: Essai de Lixiviation X 31-210, 1988, Association Française de Normalisation (AFNOR), Paris.

24. Bericht zum Entwurf für eine technische Verordenung über Abfälle (TVA), 1988, Département Fédéral de l'Intérieur. Switzerland.

25. USEPA – Landfill Disposal Restrictions: 62 FR 41005, July 31, 1997 and 62 FR 63458, December 1, 1997.
26. Characterisation of Waste – methodology guideline for the determination of the leaching behaviour of waste under specified conditions. PrENV 12920, CEN/TC 292, CEN (1996).
27. Characterisation Leaching Test – percolation test – upflow. CEN/TC 292 Working Group 6, work item 292016, formal vote 2002.
28. Characterisation of Waste – leaching behaviour test – influence of pH on leaching with initial acid/base addition, CEN/TC 292 Working Group 6, CEN enquiry 2001.
29. Characterisation of waste – leaching behaviour test – influence of pH on leaching with continuous pH control. CEN/TC 292 Working Group 6 in preparation, CEN enquiry 2002.
30. Characterisation of waste – leaching behaviour test – Dynamic leach test for monolithic materials. CEN/TC 292 Working Group 6, work item in preparation, 2001.
31. Council Directive 1999/31/EC of April 26, 1999 on the landfill of waste. DGXI, 16-7-1999, L182/1 – 19.
32. Ph. Quevauviller, W. Cofino, R. Cornelis, P. Fernandez, R. Morabito and H.A. van der Sloot, *Intern. J. Environ. Anal. Chem.* 1997, **67**, 173.
33. O. Hjelmar, *Waste Manag. Res.*, 1990, **8**, 429.
34. S. Ferrari, Chemische Charakterisierung des Kohlenstoffes in Rückständen von Mühlverbrennungsanlagen: Methoden und Anwendungen, Dissertation ETH Nr. 12200, Zürich, 1997, 128.
35. NEN 7341, Determination of the availability for leaching from granular and monolithic contruction materials and waste materials, Sept. 1993.
36. H.A. van der Sloot, R.P.J.J. Rietra, R.C. Vroon, H. Scharff and J.A.Woelders, 'Similarities in the Long Term Leaching Behaviour of Predominantly Inorganic Waste, MSWI Bottom Ash, Degraded MSW and Bioreactor Residues', *Proceedings of the 8th Waste Management and Landfill Symposium*, T.H. Christensen, R. Cossu and R. Stegmann (eds.), Vol I , 2001, 199–208.
37. R. Cossu, R. Laraia, F. Adani and R. Raga, 'Test Methods for the Characterisation of Biological Stability of Pretreated MSW in Compliance with EU Directives', *Proceedings of the 8th Waste Management and Landfill Symposium*, T.H. Christensen, R. Cossu and R. Stegmann (eds.), Vol I , 2001, 545–554.
38. R.N.J. Comans, J. Filius (LUW), A. van Zomeren and H.A. van der Sloot, Haalbaarheidsstudie naar genormaliseerde methoden voor de bepaling van de beschikbaarheid en rol van organische stof in grond, afval- en bouwstoffen m.b.t. verhoging van de uitloging van slecht water-oplosbare verontreinigingen, ECN-C-00-060, 2000.
39. D.G. Kinniburgh, W.H. van Riemdijk, L.K. Koopal, M. Borkovec, M.F. Benedetti and M.J. Avena, 'Ion Binding to Natural Organic Matter: Competition, Heterogeneity, Stoichiometry and Thermodynamic Consistency', *Colloids and Surf. A.*, 1999, 147–166.
40. M.F. Benedetti, W.H. van Riemdijk, L.K. Koopal, D.G. Kinniburgh, D. Gooddy and C.J. Milne, *Geochim. Cosmochim. Acta.*, 1996, **60**, 2503.
41. Environment Canada, *Compendium of Waste Leaching Tests*, Environmental Protection series, Report EPS 3/HA/7, 1990.
42. S.M. Wallis, P.E. Scott and S. Waring, Review of Leaching Test Protocols with a View to Developing an Accelerated Anearobic Leaching Test. AEA-EE-0392, Environment Safety Centre, 1992.
43. H.A. van der Sloot, 'European Activities on Harmonisation of Leaching/Extraction Tests and Standardisation in Relation to the Use of Alternative Materials in Con-

struction', *ICMAT 2001 Symposium on Advances in Environmental Materials*, Singapore, July 1-6, 2001.

44. D.S. Kosson, H.A. van der Sloot, F. Sanchez and A.C. Garrabrants, 'An Integrated Framework for Evaluating Leaching in Waste Management and Utilisation of Secondary Materials', USEPA, in preparation, 2001.

45. H.A. van der Sloot, R.P.J.J. Rietra, D. Hoede, Evaluation of Leaching Behaviour of Selected Wastes Designated as Hazardous by Means of Basic Characterisation Tests, ECN-C-00-050, 2000.

46. V.J.G. Houba and I. Novozamsky, *Fres. J. Anal. Chem.*, 1998, **360**, 362.

47. Compliance Leach Test CEN TC 292 Working Group 2 (1996): 'Characterisation of waste. Leaching. Compliance test for leaching of granular waste materials. Determination of the leaching of constituents from granular waste materials and sludges', Draft European Standard, prEN 12457, Parts 1-4.

48. S.K. Gupta and C. Aten, *Inter. J. Environ. Anal. Chem.*, 1993, **51**, 65.

49. A.M. Ure, R. Thomas and D. Litlejohn, *J. Environ. Anal. Chem.*, 1993, **51**, 25.

50. G. Rauret, J.F. López-Sánchez, A. Sahuquillo, R. Rubio, C. M. Davison, A.M. Ure, Ph. Quevauviller, 'Improvement of the BCR Three Step Sequential Extraction Procedure Prior to the Certification of New Sediment and Soil Reference Materials', *J. Environ. Monit.*, 1999, **1**, 57.

51. I. Novozamski, Th. M. lexmond and V.J.G. Houba, *J. Environ. Anal. Chem.*, 1993, **51**, 47.

52. Dynamic Leach Test for Monolithic Waste Material, New work item of CEN TC 292, Working Group 6, 2001.

53. J.D. Allison, D.S. Brown, K.J. Novo-Gradac and C.J. Minte, QA2 /PRODEFA2: A Geochemical Assessment Model for Environmental Systems: Version 3.0 User's Manual, EPA/600/3-91/021, US-Environmental Protection Agency, Athens GA 30613, 1991, 106.

54. M.G. Keizer and W.H. van Riemsdijk, ECOSAT, Department of Environmental Sciences, Subdepartment Soil Science and Plant Nutrition, Wageningen Agricultural Univ., Netherlands, 1998.

55. J.J. Dijkstra, H.A. van der Sloot and R.N.J. Comans, 'Process Identification and Model Development of Contaminant Transport in MSWI Bottom Ash', *Proceedings of WASCON 2000*, Harrowgate, UK, May/June 2000, in press.

56. J.A. Meima and R.N.J. Comans, *Env. Sci. Tech.*, 1999, **31**, 1269.

57. A.-M. Fällman, 'Leaching of Chromium from Steel Slag in Laboratory and Field Tests – a Solubility Controlled Process', in *Waste Materials In Construction – Putting Theory into Practice*, J.J.J.M. Goumans, G.J. Senden, H.A. van der Sloot (eds.), Studies in Environmental Science 71, Elsevier Science Publishers, Amsterdam, 1997, 531–540.

58. L.K. Koopal, W.H. van Riemsdijk, J.C.M. de Wit and M.F. Benedetti, *J. Colloid Interface Sci.*, 1994, **1666**, 51.

59. M.F. Benedetti, C.J. Milne, D.G. Kinniburgh, W.H. van Riemdijk, L.K. Koopal, *Environ. Sci. Technol.*, 1995, **29**, 446.

60. M.F. Benedetti, W.H. van Riemdijk and L.K. Koopal, *Environ. Sci. Technol.*, 1996, **30**, 1805.

61. J.C.M. De Wit, W.H. van Riemsdijk and L.K. Koopal, *Env. Sci. Technol.*, 1993, **27**, 2005.

62. M.J. Avena, L.K. Koopal and W.H. van Riemsdijk, *J. Coll. Inter. Sci.*, 1999, **217**, 37.

63. D.G. Kinniburgh, *Fit User Guide*, British Geological Survey, Technical Report WD/93/23.

64. Ph. Quevauviller, G. Rauret, R. Rubio, J.-F. López-Sánchez, A. Ure, J. Bacon and H. Muntau, *Fresenius J. Anal. Chem.*, 1997, **357**, 611.
65. *Euro-soils Identification, Collection, Treatment and Characterisation*, Special publication No. 1.94.60, Environment Institute, JRC, 1994.
66. Validation of CEN/TC 292 Leaching Tests and Eluate Analysis Methods EN-12457 (parts 1–4), EN 13370 and EN 12506 in cooperation with CEN/TC 308, Work Program, June 28th, 1999.
67. Ph. Quevauviller, *Trends Anal. Chem.*, 1998, **17**, 632.
68. H.A. van der Sloot, D.S. Kosson, T.T. Eighmy, R.N.J. Comans and O. Hjelmar, in *Environmental Aspects of Construction with Waste Materials*, J.J.J.M. Goumans, H.A. van der Sloot, Th.G. Aalbers (eds.), Studies in Environmental Science 60, Elsevier Science Publishers, Amsterdam, 1994, 453–466.
69. D.S. Kosson and H.A. Van der Sloot, in *Waste Materials In Construction – Putting Theory into Practice*, J.J.J.M. Goumans, G.J. Senden and H.A. van der Sloot, (eds.), Studies in Environmental Science 71, Elsevier Science Publishers, Amsterdam, 1997, 201–216.
70. EU Sludge Directive 86/278/EEC, up for revision, draft April 2000.
71. Working Document on Biowaste, EU DG Environment, February 2001.
72. H. Langenkamp and L. Marmo, Workshop on Harmonisation of Sampling and Analysis Methods for Heavy Metals, Organic Pollutants and Pathogens in Soil and Sludge, 8–9 February 2001, Stresa, Italy.

# Abbreviations

| | |
|---|---|
| CEN | European standardisation organisation |
| ISO | International standardisation organisation |
| DIN | German standardisation organisation |
| AFNOR | French standardisation organisation |
| NEN | Dutch standardisation organisation |
| EN | European standard |
| ENV | European experimental standard |
| TC | Technical Committee |
| WG | Working group |
| L/S | liquid to solid ratio (L kg$^{-1}$) |
| DOC | dissolved organic carbon |
| TOC | total organic carbon |
| LOI | loss on ignition |
| EPTOX | Environmental Protection Agency Toxicity test (USA) |
| NICCA | Non-ideal Consistent Competitive Adsorption |

# Laboratories Participating in the Various Reference Material Certification Studies (Chapters 2, 3, 5)

## Austria
Bundesamt und Forschungszentrum für Landwirtschaft, Vienna

## Belgium
Universiteit Gent, Laboratory of Analytical and Agro-Chemistry, Ghent

## Denmark
Teknologisk Institut, Kemiteknik, Taastrup

## Finland
Agricultural Research Centre, Jokioinen
Geological Survey of Finland, Espoo

## France
Institut National d'Agronomie, Laboratoire de Chimie Analytique, Paris
Institut National de Recherche Agronomique, Arras
Institut National de Recherche Agronomique, Villenave d'Ornon
Université de Montpellier I, Montpellier
CEMAGREF, Lyon
Laboratoire Central des Ponts et Chaussées, Bouguenais

# Germany

Bundesanstalt für Materialforschung und- prüfung, Berlin
Federal Research Centre of Agriculture, Braunschweig-Volkenrode
Institut für Wasser, Boden und Lufthygiene, Berlin
Metaleurop, Leiter Labor, Arbeit- und Umweltschutz, Nordenham
Technische Universität Hamburg-Harburg, Hamburg

# Greece

Aristotelian University, Laboratory of Analytical Chemistry, Thessaloniki

# Ireland

Agriculture and Food Development Authority, Wexford

# Italy

Università di Bari, Istituto di Chimica Agraria, Bari
Università di Udine, Dipartimento du Chimica, Udine

# Portugal

Estação Agronómica Nacional, Oeiras
Universidade Nova, Disciplinas de Ecologia Aquática, Lisboa

# Spain

Estación Experimental del Zaidin, Granada
Universidad de Huelva, Departamento de Química Analítica, Huelva
Universidad de Barcelona, Departamento de Química Analítica, Barcelona
Universidad de Córdoba, Córdoba

# Sweden

Chalmers University of Technology, Gothenburg
University of Uppsala, Erken Laboratory, Uppsala

# Switzerland

Station Fédérale de Recherches en Chimie Agricole, Liebefeld-Bern

# The Netherlands

Landbouw Universiteit, Wageningen
Research Institute for Agrobiology, Haren

# United Kingdom

The Macaulay Land Use Research Institute, Aberdeen
University of Reading, Department of Soil Sciences, Reading
University of Strathclyde, Department of Pure and Applied Chemistry, Glasgow

# European Commission

Joint Research Centre, Environment Institute, Ispra, Italy

# Subject Index

www.ingramcontent.com/pod-product-compliance
Lightning Source LLC
Chambersburg PA
CBHW031953180326
41458CB00006B/1697